DARWIN SLEPT HERE

DARWIN SLEPT HERE

DISCOVERY, ADVENTURE, AND SWIMMING IGUANAS IN CHARLES DARWIN'S SOUTH AMERICA

ERIC SIMONS

THE OVERLOOK PRESS
WOODSTOCK & NEW YORK

For my parents

This edition first published in the United States in 2009 by
The Overlook Press, Peter Mayer Publishers, Inc.
Woodstock & New York

WOODSTOCK:
One Overlook Drive
Woodstock, NY 12498
www.overlookpress.com
[for individual orders, bulk and special sales, contact our Woodstock office]

NEW YORK:
141 Wooster Street
New York, NY 10012

Cataloging-in-Publication Data is available from the Library of Congress

Book design and type formatting by Bernard Schleifer
Manufactured in the United States of America
ISBN 978-1-59020-220-3
FIRST EDITION
1 2 3 4 5 6 7 8 9 10

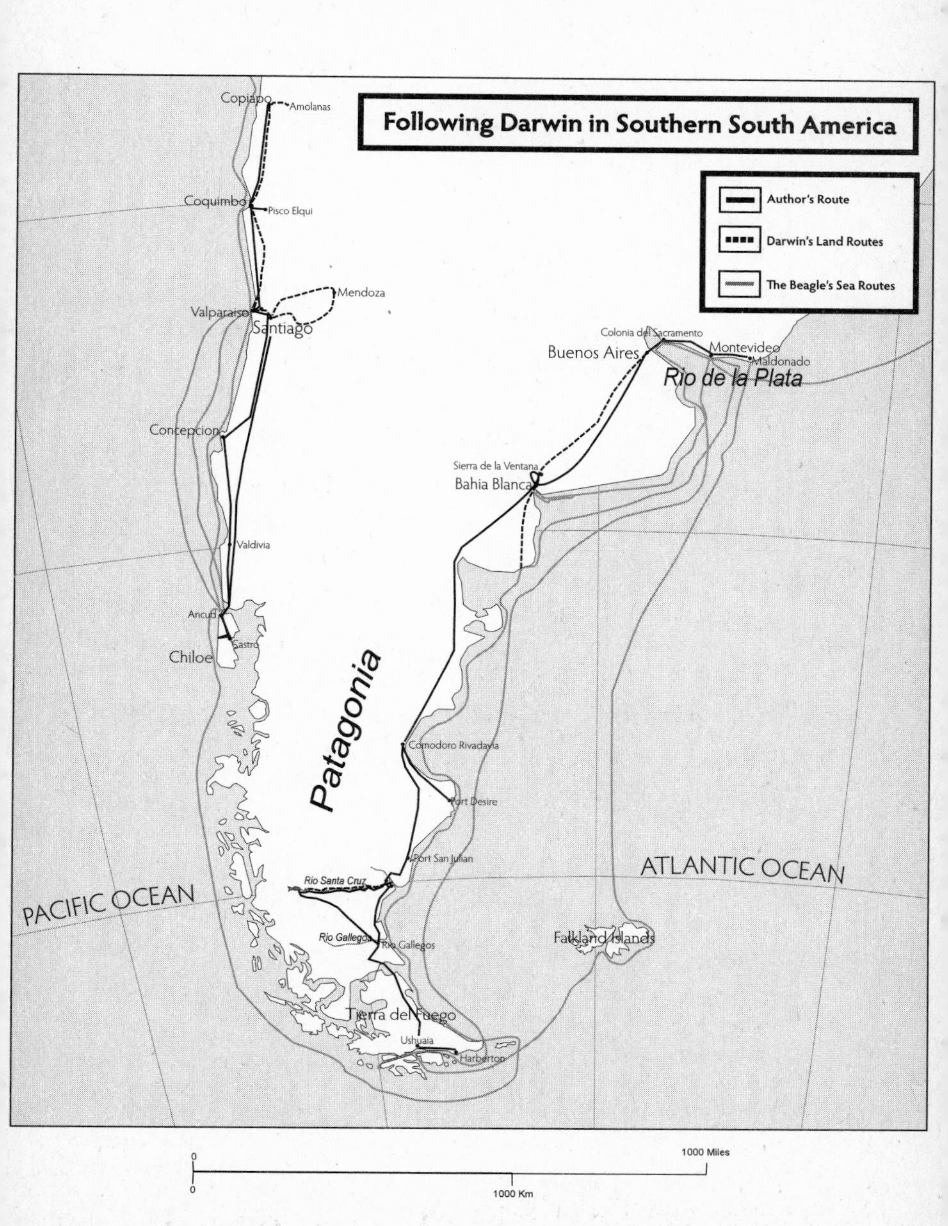

Following Darwin in Southern South America
Author's Route
Darwin's Land Routes
The Beagle's Sea Routes
Copiapo
Amolanas
Coquimbo
Pisco Elqui
Mendoza
Valparaiso
Santiago
Colonia del Sacramento
Buenos Aires
Montevideo
Maldonado
Rio de la Plata
Concepcion
Sierra de la Ventana
Bahia Blanca
Valdivia
Ancud
Castro
Chiloe
Patagonia
Comodoro Rivadavia
Port Desire
Port San Julian
ATLANTIC OCEAN
Rio Santa Cruz
PACIFIC OCEAN
Rio Gallegos
Rio Gallegos
Falkland Islands
Tierra del Fuego
Ushuaia
Harberton
0
1000 Miles
0
1000 Km

CONTENTS

PART III: DISCOVERY

INTRODUCTION
The World's Most Famous Iguana Hurler

> *Happiest at home with his notebooks and his microscopes.*
>
> —INTRODUCTION TO AN EXHIBIT OF DARWIN ARTIFACTS
> NEW YORK MUSEUM OF NATURAL HISTORY, MAY 2006

EVOLUTION HAD DONE THE THING RIGHT. The marine iguana of the Galapagos Islands swam well. Dined well. Lounged well. It basked in the sun, it munched seaweed, it strutted out for an occasional constitution-improving swim, all until one cloudless, sweltering September afternoon in 1835, when a young man stepped ashore and ruined everything.

Charles Darwin had not yet conceived the theory of evolution by natural selection. Five months shy of his twenty-seventh birthday, tall and thin and already distinctively heavy browed, he had not yet acquired a reputation as a scientist, had not yet published a celebrated travelogue about South America (or an influential treatise on tropical corals), and had not yet had a species of ostrich named after him. His visit to the Galapagos came at the tail end of a five-year trip around the world, and it did not act on him as one of those Sistine-Chapel-ceiling, hand-meets-hand kind of moments. But Darwin was in the midst of a travel-induced transformation, combining his childhood love of exploration and biology with an increasingly sophisticated ability to catalogue nature. When he published *The Origin of Species* twenty-four years later, it was notable for the meticulous observational detail Darwin used to support his theory. For someone who delighted in scientific inquiry, the reptilian megafauna swarming the Galapagos was a scaly, ugly, crawling—and terrific—learning opportunity.

Darwin spent one day studying tortoises, chasing them, riding them, and upending them to see if they could right themselves. He spent another day with the marine iguanas, and it was not a good day to be a member of the lizard kingdom. He cut up the iguanas to see what they were eating (seaweed), and in his journal, he disparaged their color ("dirty black"), their disposition ("stupid and sluggish"), and their looks ("hideous"). He and a co-conspirator tied one animal to a rock and dropped it off their boat, the Beagle, to see what would happen ("when, an hour afterwards, he drew up the line, it was quite active"). He also noticed that some of the iguanas seemed to like the water, and he wondered: How well did they swim?

On the morning that Darwin chose to answer this question, it became evident that in one way, at least, evolution had failed the iguana: It had given it no recourse at all for dealing with thrill-seeking British naturalists. Darwin strode across the craggy rocks toward a napping "imp of darkness," cornered it, snatched it by the tail, and hurled it into a pool left by the receding tide. The iguana, no doubt wondering what had gone wrong on a day that had started so pleasantly, swam straight back to its sunning rock.

Charles Darwin was a scientist at heart, and a good scientist always repeats his experiment. As the aggrieved beast climbed dripping from the pool, Darwin jumped forward again, clasped the iguana firmly in hand, and drew back. And then, in the name of science, discovery, and swimming iguanas, he hurled it into the sea.

In his five-year stint as geologist, naturalist, and traveler on the HMS Beagle, a gloriously happy Darwin galloped with Patagonian gauchos, stormed Montevideo armed with cutlasses and a knife between his teeth, beat his bare chest to properly greet an indigenous man in Tierra del Fuego, chased exotic birds, beetles, and butterflies through the jungles of Brazil, and, of course, took up iguana-tossing. He lived the swashbuckling explorer's life that modern travelers desire and almost never achieve, and he penned the book detailing the places and adventures that travelers hope to experience. A sense of exhilaration pervades *The Voyage of the Beagle*—exhilaration totally dissonant with the Charles Darwin we remember today as a wrinkled, heavy-eyebrowed, white-bearded, finch-beak measurer. In the United States, at least, Darwin exists as an almost mythical figure: praised or reviled, overemphasized or over-criticized, beyond personality and beyond the affable cheeriness that defined most of his life. Critical creationists want to see Darwin as an old tormented evolutionist, unhappy with his work and shut up in his country house in England. Many evolutionists want to see him as a rigid scientist, a pristine visionary beyond the petty events of daily life. Biographers tend to approach the young, happy, excitable Darwin as merely a training ground for what would come later, picking apart his journals for clues to his thinking on evolution and discarding many of his early energetic adventures.

Darwin never forgot these experiences. "The voyage of the Beagle has been by far the most important event in my life and has determined my whole career," he wrote in his autobiography. "I feel sure that it was this training which has enabled me to do whatever I have done in science."

The science was pretty much the extent of what I knew until I happened across *The Voyage of the Beagle* for the first time, and read about the iguana-lobbing, and immediately went green with nineteenth century naturalist envy. The guy got to chase, catch, and throw iguanas—repeatedly—and call it research! But I'd seen the pictures—

Darwin was some old tormented guy with facial fungus like Santa Claus. Well, how the hell did *that* happen?

My introductory encounter with *The Voyage of the Beagle* came in the late afternoon of a snowy day in Ushuaia, Argentina, the self-proclaimed southernmost city in the world. I was there in the very last days of a long trip, trying to put as much distance as possible between myself and a frustrating post-college stab at employment. Trying, in fact, to put seven thousand, one hundred, forty-three-point-four miles between us—I really loathed that job. For two years I'd worked the 4 P.M.–1 A.M. shift as a newspaper copy editor—the most soul-leeching job in an industry that operated on the principle that the more leeches dangling from employees the better. On nights when I couldn't indulge my primary form of relaxation (hurling dirt clods at the "SLOW" sign in the parking lot), I turned to a secondary method of stress relief: thumbing through the atlas. Retreating into the corner of my cubicle, I would dream of faraway lands, of reaching that glorious level of understanding where the Brazilian forest was more than little orange lines and squiggly blue rivers on a map, but a vast, green, three-dimensional place full of real people living real lives and chasing real lizards. So one miserable night—probably as I was bumping the headline size on a car crash story—I bought plane tickets. Which was how I found myself a few months later, idling through my savings account in Tierra del Fuego, and stumbling into a bookstore that stocked English-language books on the same afternoon a pair of friends and I had bounded through snowy mountains overlooking the Beagle Channel.

I decided on impulse to learn more about the ship that gave its name to the channel, and that provided Darwin with lodging for nearly five years. That evening I curled up in the hostel common room, watched snow swirling over the channel, and started to read—and kept reading for the next two weeks as I stumbled back to Santiago, Chile, and returned home. But I could not leave that image of Darwin, and the snow falling over the Beagle Channel, behind. Back in California, with no burning ambition to return to graveyard-shift proofreading, I started to map out a new voyage.

The Beagle, a surveying ship under the command of Captain Robert FitzRoy, left England in December of 1831 and arrived on the east coast of South America, in Brazil, in February of 1832. The ship spent five months in Brazil, including a three-month stop in Rio de Janeiro (Darwin rented a cottage and lived onshore), then headed south down the coast of Brazil to the world's widest estuary, the Rio de la Plata, between present-day Argentina and Uruguay. From there, FitzRoy launched multiple surveying trips, heading south to Patagonia and Tierra del Fuego before returning to Montevideo or Buenos Aires. Darwin, who suffered from terrible seasickness ("the misery is excessive and far exceeds what a person would suppose," he complained early on), would leave the boat whenever possible to make land expeditions. These weren't always daytrips—Darwin would sometimes quit the Beagle for weeks at a time and cover a huge distance over harsh terrain. Twice he exceeded four hundred miles, including a two-week ride with a band of gauchos from Northern Patagonia to Buenos Aires that left him hugely enamored of the nomadic lifestyle and idly speculating about the health benefits of his new all-red-meat diet. ("I found this new regimen agreed very well with me," Darwin wrote, "but I at the same time felt hard exercise was necessary to make it do so.")

In June 1834, after two years spent exploring back and forth between the Plata and Tierra del Fuego, the Beagle rounded the southern tip of the continent and started surveying the west coast of South America. Over the next year the ship moved gradually northward, while Darwin made land expeditions into the interior to inspect Chilean mines or examine the geology of the Andes. On his final—and longest—land journey, he rode north 420 miles from Valparaíso into the world's driest desert and met up with the Beagle again two months later near the tiny mining town of Copiapó, in northern Chile. From that point the ship continued north without backtracking and reached Lima, Peru, in July 1835. (Darwin didn't explore around Lima because of political trouble.) From Lima the Beagle headed west to the Galapagos Islands, where it stayed for a month, and then on through the Pacific—with stops in Tahiti, New Zealand, and Australia. ("Australia is rising, or indeed may be said to have risen, into a grand centre of civilization, which, at some not very remote period, will rule as empress over the southern hemisphere," Darwin proclaimed. "To hoist the British flag seems to draw with it as a certain consequence, wealth, prosperity, and civilization.") The ship sailed past Africa and rode the wind all the way back across the Atlantic to northern Brazil, returning to Darwin's first landing-spot in South America. After a few days in Bahia, which Darwin spent reexamining the Brazilian jungle, the Beagle departed for home, and Darwin arrived back at his family house in October 1836.

During the trip, Darwin kept three journals, two scientific and one personal. He wrote in them almost every day, filling the personal diary with his adventures and observations and the scientific ones with descriptions of plants, animals, fossils, and geology. He took any opportunity to mail his completed work home, and when he returned to England after nearly five years on the Beagle, he published a combination of his two journals as a book titled *The Voyage of the Beagle: Journal of Researches Into the Natural History and Geology of the Countries Visited During the Voyage of H.M.S. Beagle Round the World*.

With its publication, Darwin became a famous and well-recognized naturalist, explorer, and author, twenty years before writing *The Origin of Species*. He could have stopped there and been remembered forever as an amiable, moderately accomplished scientist and left the evolution-related mud-slinging for Gregor Mendel, the father of modern genetics, or Alfred Russel Wallace, the scientist who came up with a theory of natural selection at the same time as Darwin. But he did not stop, and so became the name and face of one side in a debate so rancorous it threatens to obliterate his real personality.

That personality shines through the pages of *The Voyage of the Beagle*. Three years after Darwin's death, his wife, Emma, picked up his journal and started reading as a way to remember her husband. "It makes me feel so happy as if I was going with him; only I want to ask him so many questions," she wrote in a letter to their son William. She had recently read Darwin's description of a brightly colored black-and-vermillion toad, which, in a reference to seventeenth century English author John Milton, he called "a fit toad to preach in the ear of Eve." Upon encountering one of these toads on a dry plain, Darwin, "thinking to give it a great treat, carried it to a pool of water; not only was the little animal unable to swim, but I think without help it would soon have been drowned."

To Emma Darwin, this incident captured her husband's charm perfectly. "The real man comes out constantly," she wrote in her letter, "e.g. thinking to give a toad in a dry place in Rio Plata 'quite a treat' by taking it to a pond & nearly drowning it."

It is neither possible nor desirable to follow the Beagle chronologically (unless you've got five years). The ship spent so much time traveling up and down the South American coast, poking into inlets and taking soundings as it went, that a journey in Darwin's exact footsteps would be tedious at best. But once the Beagle passed Cape Horn and started to explore the west coast of Chile, it didn't return to the east coast (except for the weeklong pit stop in Bahia on its way back to England). So I conceived of a journey in two parts, one visiting Darwin highlights on the east coast—from Brazil in the north to

Tierra del Fuego in the south—the other tracing the Beagle on the west coast, from southern Chile to the northern desert.

But there's more to following Darwin than just standing where he stood. One of the things that most impressed me about *The Voyage of the Beagle* was how much Darwin sounded like twenty-somethings swapping stories in a h o stel bunk room. I hoped that a return visit to South America, this time with Darwin's diary in hand, would deepen my understanding of the legends and people that fascinated Darwin when he wasn't working. I hoped I'd be able to trace the historical roots of some of South America's most persistent travelers' tales, like the challenging Brazilian forests and culture and the romantic allure of Patagonian plains, while peering into the cultural identity of the continent's most enduring places.

I also hoped that setting off on such a trip while I, too, was in my mid-twenties might allow me easier access to Darwin's emotions while observing a continent amid fitful, often violent, transformation. Even in the 1830s, the places the Beagle dropped anchor were in the midst of a transformation—facing floods of immigrants, the rotten last years of colonial governments, expansion of grazing land and mining, the resulting pressure on traditional nature and culture, and wars between settlers and indigenous people. Darwin was tossed into the middle of this with little context from his English education. Now, as these same areas face new pressures—globalization, economic woes, environmental degradation—and with my own education having offered little in the way of background, I hoped to employ the same curiosity to better understand what, exactly, is at risk.

It's hard not to be sentimental. In 1835, you could toss a Galapagos iguana around a few times to see what would happen. Right now, you'd leap for the tail and probably end up with a restraining order. In Darwin's era, rolling boulders off a cliff to see how much noise they'd make was high sport. Right now, touching the sacred parts of a national park and disturbing others' wilderness tranquility gets you raised eyebrows and a ranger-escort to the front gate. Even in a generation, things have changed d r amatically in South America. Water is dirtier. Slums are growing. Habitats are disappearing. Iguanas, tortoises, fish, and just about every other kind of flora and fauna are declining.

The early twenties are a time for worrying about risk, loss, and the past. Little wonder, I think, since when you finish college you suddenly realize the huge number of things you could be doing with your life. Like Darwin, a lot of twenty-somethings simultaneously calculate the trajectory of the safe route, and then run off screaming for variety. My own quarter-life crisis wasn't just a panic about choosing between options, it was a panic about when or even if I might have that variety of choice again. I grew up during perhaps the most pampered, materialistic time in American history, and then hit adulthood and realized that my generation was going to make less money than our parents did, with fewer work opportunities for lower pay, and that I live in a world in which the United States is not the world's only superpower, with looming repercussions for deferred decisions on climate change, Social Security, and tax cuts. That's why decision-making morphs so easily into fretting about losing our youth while we are, in fact, living it. I'm told this mania wanes when you hit thirty, and it certainly did for Darwin, who, upon touching land in England again at age twenty-seven, had permanently cured his wanderlust.

A note on organization: Rather than belabor all the overnight bus rides it takes to actually follow Darwin around (although I should mention that overnight buses in Argentina are sweet—roomy reclining chairs, snacks, movies—while buses in Brazil are scary—crazy drivers, frequent crashes, and movies like *The Matrix*, full of noise and mayhem), I've split the story into three thematic parts. The first looks at how Darwin the naturalist and geologist saw South America's east coast, from the jungle of Brazil to the plains of Patagonia. The second tracks his forays into politics and culture as he railed against slavery, rode with gauchos, and studied Native Americans. The final part follows Darwin up the west coast in a more chronological fashion, taking into account what he'd learned and how he'd changed, and leading northward into the Chilean desert where both he and I chose to end our overland travels.

In picking Darwin sites to visit, I tried to look for significance beyond science. To compare modern Rio de Janeiro to the green world that inspired Darwin to write, in his autobiography, that nothing can "exceed in sublimity the primeval forests undefaced by the hand of man." To track the fate of the captured natives of Tierra del Fuego who rode along with the Beagle, pitied Darwin's seasickness, and were repatriated to their homeland. To take in summit views across the continent to see how many of them cued up for me, as they did for Darwin, the opening choral strains of Handel's Messiah. And, with any luck, to find the twenty-first century equivalent of a day spent lobbing marine iguanas.

PART I
EXPLORATION
THE CHANGING LAND ON THE EAST COAST

1. BRAZIL
A Chaos of Delights

> *I do not know what epithet such scenery deserves: beaut iful is much too tame; every form, every colour is such a complete exaggeration of what one has ever beheld before. If it may be so compared, it is like one of the gayest scenes in the Opera House or Theatre.*
>
> —BEAGLE DIARY, JUNE 1, 1832

BRAZIL IS THE KIND OF PLACE where you feel something's happening that's absolutely delightful and fascinating and completely foreign to everything you know, only you can't quite grasp what it is. That weird paranoia is amplified by the stunningly disorienting scenery of places like Rio de Janeiro, where huge, smooth cones of rock swoop up and down, clad in a thriving blend of palms, ferns, cactuses, and scrubby bushes, while the glittering jumble of city runs right up to—even into—famous mountain landmarks rising out of the water.

I started my voyage with Brazil because that's where Darwin started his and because it was that exhilarating landscape that stood out most in his mind as he reflected on his trip later, as an old man. Also, as I soon found out, there was an easy comparison to make between Darwin's take on the new nature around him and my own adrenaline-fueled excitement at Brazil's crazily, incomprehensibly, wonderfully different city landscape.

The feeling I was missing something only increased as I tried to adjust to Brazilian social life. A functional grasp of Spanish couldn't get me beyond "Hello, how are you?, The dog is green" conversations in Portuguese. I was bunking down in a cramped youth hostel in Botafogo and had trouble connecting with my hostel mates, who tended to divide into camps of 1) fluent Portuguese speakers from Europe and North America who had immersed themselves in Brazilian life and did things like teaching in schools and looking down upon us ignorant gringos, 1a) not-yet-fluent Portuguese speakers who were dating Brazilians and looked down on us ignorant gringos, and 2) blissfully ignorant gringos whose knowledge of Brazil started and stopped with an aesthetic appreciation of the thong bikini.

For dorm roommates I had drawn a preening, chiseled German and an aloof Frenchman. The German liked to strut around the room in a Speedo and, while admiring his rippling bronzed abs, lecture us about the club scene in heavily accented Tarzan-style English. "So there is this place I went to last night where the cover was 120 reales," he told us. The Frenchman raised an eyebrow at the cost, to which the German quickly added, "But you can drink up to 100 of it. And it is worth it just to see it. They have the most beautiful girls there."

The floor near the German's bed was littered with wrinkled-up napkins used to capture names and phone numbers. "Flavia," whose napkin had blown over in my direction and lay face up on the floor, had taken the time to write hearts around her number.

When the German took an interest in what I was doing in Brazil, however, I left him baffled by saying I wanted to go to the Tijuca National Park.

"What's that?"

"Hiking," I said. "In the rainforest, in the mountains. Want to come?"

"Oh no," he said. "Not for me. I'm going to the beach again. Have you seen the girls on the beach?" He clicked his tongue and gave me a thumbs-up.

Both roommates were asleep—Tarzan sprawled out in his briefs, sheet pushed away—when I tiptoed out early on a gray, humid Thursday morning. Botafogo was a one-time suburb of Rio where Darwin had taken a cottage to use as a peaceful base for trips into the surrounding jungle, and it had since vanished—along with Rio's other suburbs—under an onslaught of humanity and cinder block. I boarded a bus headed north, to Tijuca, where I expected to find not just a Darwin site, but a remnant of tropical forest, and a much-needed dose of natural tranquility. We lurched and honked our way along in a cloud of exhaust, past vendors and flapping flags emblazoned with the Brazilian national motto, "Order and Progress."

Order did not exist here the way it does in other countries. Sidewalks all over town blinked in and out of existence like a demonstration of quantum theory, ending abruptly, sometimes starting again a few blocks later, but certainly following no predictable pattern. Houses of various and dissonant architectural styles leaned on one another. Glass medical centers clung to the gabled offices of massage therapists, which cast their shadows on brown adobe liquor stores. Street vendors hawked batteries, scissors, pirated DVDs, cell phones, peanuts, coconuts, electric sanders, and leather-covered steering wheels, in case drivers wanted to pimp their ride mid-commute. And then the Brazilians—tanned, lean people in shorts, T-shirts, and flip-flops everywhere, all on their way somewhere. (I suppose, given the national motto, they were on their path to progress.) The hum of conversation mixed with the cries of the vendors, the rumble of idling motorcycles, the drone of airplanes and sightseeing helicopters, and the whines and sighs of trucks braking to stops and moving on again. Blasts of exhaust, sweat, grime. and dust joined the humidity to create a suffocating cloud. In the shade of the omnipresent buses, manic taxis, scooters, policemen, pushcarts, and bicycle delivery riders for "Bob's Burgers" competed for space on the street. One messenger company that dispatched riders to sites around the city was called, in English, "Boy Delivery."

While I struggled with a radically different version of social order, Darwin confronted in Brazil a radically different version of the natural world. Darwin hailed from a country that, after thousands of years of cultured habitation, had essentially become one big garden. A big cultivated garden in a cold climate. A two-month boat ride and he found himself sweating and marveling at an intensity of color he had never encountered outdoors before. Vegetation grew uncontrollably, in a way that seemed a direct challenge to a stiff-upper-lip-loving Englishman's sense of the universe, and his reaction to virgin forest pretty much mirrored my German roommate's reaction to the girls on the beach: "Delight is however a weak term for such transports of pleasure" and "I can only add raptures to the former raptures." He concluded his diary entry from his second day in Brazil with "Full of enjoyment one fervently desires to live in retirement in this new & grander world."

In the 1830s, Rio de Janeiro amounted to a whole lot of neat monolithic rock formations with jungle covering everything in between. Twelve million people later, the city forced its will on what little natural spaces remained, presenting its own profusion of growth—an urban mirror to the bursting vegetation of Darwin's day. Now tall, clean skyscrapers rose above all else, lifeless towers of cement and steel punctuating the swarming, sweltering, steamy mess below, competing for attention with forest-clad mountain peaks in the background. It was the pure state of nature, still—just the "nasty, brutish and short" Hobbesian version.

After my forest-bound bus had lurched through that tumult for more than an hour, crammed streets gave way to a shadier, quieter neighborhood. The road climbed a winding hill, with thick tropical foliage hanging over the road. Gated driveways led to fantastic houses overlooking the skyscrapers and curving white sand beaches below. ("On the road, the scenery was very beautiful," Darwin declared from the same spot, "especially the distant view of Rio.") At the top of the hill, in a tranquil neighborhood called Alto da Boa Vista, another long gated driveway curled into the Tijuca National Park like the entrance to a wonderful tropical mansion. A creek and verdant forest, infused with the pure smells of tropical flowers and wet soil, quickly surrounded everything.

A guide introduced himself as Jean Marx Muñiz Belvedere and politely broke into my reverie to invite me on a walk through the forest. To get him talking in English, which he claimed to be uncomfortable with, I asked what he had done before he came to the park. "I was an artist," he said. "How you say, *trapezista*?"

He swung his arms and flashed a sly half-grin.

"A trapeze artist?" I said.

"Yes," he said matter-of-factly. "That's right."

To him, this was perfectly normal. This was one of those times when I felt I was missing something. When you finally do find someone to talk to in Brazil, he makes it sound like everyone local just happened to have a past that involved circus performances.

Jean the former trapeze-artist looked stereotypically Brazilian—tanned olive skin, hirsute, toned limbs, a head of short, dark, curly hair, and a perpetual humid-jungle fog on his thin-rimmed glasses. I asked what he knew about Charles Darwin, and he asked me how much time I had for the answer.

It turned out he didn't ask because he was a Darwin expert, but because the best thing he could think of to do with my question was to take me hiking to the highest peak in the park, the 3,400-foot-tall Pico Tijuca. "It's where we traditionally take visitors," he said. "The ones who can walk. You can walk?"

There wasn't much in Darwin's own account to follow. On June 16, 1832, he left his cottage in Botafogo early in the morning to see the waterfalls in "Tijeuka." "Neither the height or the body of water is anything very imposing," he wrote, "but they are rendered beautiful, by the dampness so increasing the vegetation, that the water appears to flow out of one forest & to be received & hidden in another below."

Now the park's largest waterfall is the 115-foot-tall Cascatinha Taunay, and the main pathway into the park crosses a bridge directly below the falls. The rainy season was coming to a close and the waterfall was flowing fast, probably more so than in Darwin's time. Moss and jungle plants clung to life on the sheer rock face supporting the falls, while vines and creepers webbed their way over the top of the waterfall. From the top, water cascaded into a small pool, flowed downhill into a larger, calmer pool, and then joined the creek running down toward the entrance to the park.

I asked Jean how the waterfall might have looked in

Darwin's time, and he surprised me. He said the park, in its present incarnation, was actually more jungly now than it was in 1832. Modern-day Parque Nacional Floresta Tijuca, Jean told me, was the result of a bold and successful environmental engineering project—one that dated to just after Darwin's visit. Rio de Janeiro's early history runs along the usual lines—backwater town booms when resources get discovered—but with a bit of a twist. The entire Portuguese royal family moved to the capital of their Brazilian territory when Napoleon invaded Portugal in 1808. The population of Rio boomed, and it kept expanding, carving rapaciously into the surrounding jungle in search of more land for plantations and suburbs. The entire forest where we stood admiring hundred-foot-tall trees that appeared ancient had been burned to the ground and turned into a coffee plantation. "Naked," Jean said, staring off into the jungle. "This entire spot was naked."

In the mid-1810s, King John, who continued living in Rio after Napoleon's defeat, began to worry that the destruction would harm the city's water supply and issued orders that the land around streams be replanted with trees to guarantee a potable water supply. His commands went ignored until the springs that provided fresh water for the city went dry and Rio suffered four massive droughts. Darwin, who visited in 1832, missed these—the last had happened in 1829, and the next would hit in 1833—and he didn't mention any water supply troubles. But the most severe dry spell, which occurred in 1844, finally convinced the government to act. By that time Brazil was independent from Portugal, and the country's new rulers decided to put some money and thought into conserving the forests and waterfalls of Tijuca. In this case, conserving meant replanting.

In 1861, Manuel Gomes Archer was named administrator of the Tijuca Forest project, and together with six slaves, he began to replant the forest. Archer had no formal training but was considered a local expert, and he decided to replant the area with native plants in roughly the same ratio he had seen in other Brazilian forests. Over the next twelve years, Archer's six slaves planted 72,000 trees. The springs, according to a report Archer prepared near the end of this administration, regained and retained their former water levels.

Jean told me that our trail was called the Major Archer trail. On the way into the park, I had seen a sculpture at the visitor's center of a man holding out a fresh, live bromeliad, and Jean explained that it honored the six slaves who returned Tijuca to the forest.

"Do you know what the word *ecologia* means?" Jean asked.

"Ecology? Sure. Why?"

"What does it mean?"

"I don't know. The study of ecosystems—plants and animals and the environment."

"No," Jean said, "what does it *mean*? What does *eco* mean?"

"Oh. I don't know. Environment?"

"*Eco* is Greek for house, home. *Logia* is Greek for study. *Ecologia*, ecology, is the study of your home. And what does *economia* mean? *Eco*—home—and *onomia* means organization. I give this talk to our volunteers the other day. I explain to them, how can you have your home organized if you do not study it?"

Jean was clearly at home in the forest. He hiked at a blistering pace, nimbly dodging fallen trees and leaping over creeks and, all the while, prattling merrily about flora and fauna and the kinds of things that might be lying in wait behind the rotting tree trunk I was using to pull myself up the muddy trail.

"Oh, yes, many snakes here," Jean said. "Only one very poisonous though."

"Oh," I said. "Ah."

"But she is—she don't like us. She knows we are more dangerous to her. She recognize the most dangerous animal on Earth." He chuckled at this line, and its English delivery, rubbing his hands in self-satisfaction.

Jean mentioned *jacarei*—crocodiles—which used to live in the swamps around Rio but now were gone except for the huge natural area in southwestern Brazil known as the Pantanal. He was sad about this. "Jacarei, they not want to eat you. Jacarei, they are good people."

And what about sharks, I wanted to know. I've long had a soft spot for the world's most feared apex predators.

"I like sharks," Jean agreed. "Sharks, they are good people too."

After about an hour, Jean called for a rest. I was panting, sopping with sweat. Jean's glasses had now fogged over completely. He indicated the ruins of a small brick house, buried in the jungle, and we wandered over there and sat down on the remains of a windowsill. Moss covered the outside, and the doors and windows were long gone. A landslide had pushed through the back window, and a banged-up refrigerator had taken root in the middle of the room. Jean told me that someone had lived in the house and had disappeared. His ghost supposedly roamed the forest. We looked out at the canopy.

"It's just a story," Jean said, misinterpreting my silence.

While he sprawled on the cool, mossy brick, I pulled out my copy of the Beagle diary and read out loud Darwin's account of the road up to Tijuca. In his typical style he had written: "As a Sultan in a Seraglio I am becoming quite hardened to beauty. It is wearisome to be in a fresh rapture at every turn of the road. And as I have before said, you must be that or nothing."

"Yes," said Jean, nodding as I explained the word rapture. "Very beautiful."

He pointed out some small lichens on a tree, little splotches of green with a bright pink outline. "They only grow where there is no pollution," he said. "The air here is good."

I told him that I felt suffocated by the city's constant commotion within hours of arriving, and mentioned my German roommate, who was no doubt at that very moment lying on the Copacabana in his Speedo, attracting girls like a Bob's burger set out for hungry seagulls.

"I understand," Jean said. "It was like that in the circus." Seer-coos, he said.

"Party all night, sleep during the day. For a long time, that was my life. Like so many people in Rio. I met this

Italian guy a few weeks ago, who had been coming here every year for the last twenty years. He had married a girl here. And he had never heard of the Tijuca Forest. He just comes here every year to go to the beach. He told me, 'I don't like it here in the forest. It's too green. I feel suffocated here.'"

We both looked up at the light filtering through the canopy, hundreds of feet above our heads, and then back at the little fresh-air lichens. "It is funny," Jean said, "how people do not perceive what is right near them." He looked up again. "I like it here. Now, I work when everyone else rests, but is OK. I could not work in some room with four walls."

We started hiking again and moved upward through the wearisome raptures of the forest for another hour or so. (With all respect to Darwin, plodding up switchbacks in 85-degree, 85-percent humidity is far more wearisome than endless greenery.) As we neared the peak, the trail leveled out into an overlook of white rooftops and the d i stant Pão de Açúcar. The Pão is one of Rio's most identifiable landmarks—the jungle-covered rock that rises from the water in postcards—and a distinctive reminder that, even if the interior jungle looks like tropical jungle elsewhere, from this overlook you are now incontrovertibly looking at Rio de Janeiro.

Of course, there are other things that can do this. "There's the Maracanã," Jean said, pointing at the world's largest soccer stadium. He was, unsurprisingly, a soccer fan. ("I taught Ronaldhino how to play," he insisted, smiling. "Everyone says he's the best player in the world. He's just the best player playing professionally.") Other than the Maracanã, he pointed out no other buildings in the city. I think he felt unsettled by the city intruding so aggressively on his forest—his home. For the last few hours he had named every plant and bird and identified every rustle in the bushes, but faced with the sprawling, honking complexity below, he went silent.

Instead, he pointed out a vulture with white wingtips, circling high over the city. "People don't like them," he said, "but they are very beautiful at flying. Scientists say that vultures fly only to find food, but I'm not sure. I believe they are flying for fun."

We pushed through a few small clumps of grass and trees and emerged atop the second-highest peak in the park, the ground falling away below us into a breathtaking 300-degree view of Rio, splayed out on the coastal plain around the glittery silver Guanabara Bay. "I could sleep up here," Jean said. "This is my favorite view in the whole park."

"The *cidade maravilhosa*," I said. Rio de Janeiro's nickname: the Marvelous City.

"Yes," Jean said. He repeated, sadly, "*Cidade maravilhosa.*"

"Sometimes," he said, "I'm not so sure. I like to come up here and think what it would be like if the Portuguese had never come." He looked away from the city, over the jungle stretching out toward the ocean to the south. He shrugged and wandered off for a minute.

"Just like everywhere," I said when he returned. "Brazil and California are similar. Many of the Indians are basically gone."

"And with them, all the knowledge," Jean added. The topic seemed to depress him. "They knew about ecology. How to organize their home."

He asked whether I wanted to climb the rest of the way to Tijuca Peak. I said yes, of course—Darwin wouldn't have settled for the second-highest peak with the highest peak in such plain sight. Its bald dome rose from a densely forested ridge connected to where we were standing. The trail wound through the jungle to the base of the summit, then turned to climb 1,020 stairs that had been cut into the rock specifically for the King of Belgium in 1921. The president of Brazil had hoped to please the king, a renowned climber, by allowing him to access the difficult summit. The king saw the stairs and took it for the novice route, so he made his own way, rock-climbing to the top.

"It was a huge shame," Jean said.

Park workers had driven iron spikes into the rock around the stairs and strung thick metal cable between the posts as a swaying handhold. The way the posts poked out at odd angles, some of them leaning temptingly over sheer cliff edge, it looked safer to try and hang onto the steps themselves. After ten minutes of climbing, we hauled ourselves up onto a small grassy knoll that marked the top and looked down through the thin layer of smog. Below us to the east, the winding Guanabara Bay glittered in the afternoon sun, and the city of Nitteroi, on the opposite shore, appeared faint in the coastal haze. To the south, dramatic posts of jungle-covered rock cut into our view of the famous white sand crescents of Copacabana and Ipanema. In the north, planes took flight from the airport, and teeming slums climbed up the hillsides until the forest took over. The view west encompassed nothing but forest, rolling, tumultuous green hills stretching into haze.

The spectacular view left me exhilarated. I thought back to Darwin, who wrote in his first letter home from Brazil, "The exquisite glorious pleasure of walking amongst such flowers, & such trees cannot be comprehended, but by those who have experienced it." The contradiction between his repeated attempts to describe the forest, and his inability to pick enough adjectives to adequately communicate it, had been much on Darwin's mind in Brazil. His rainforest accounts were heavily influenced by a travelogue written just before his birth by the famed wandering German Alexander von Humboldt, a man who liked his descriptors. (Darwin really took to him, writing at one point, "He like another sun illumines all I behold.")

Darwin's infatuation with von Humboldt lasted for several months and extended to his writing style, so that at one point his oh-so-very-English sister wrote him a letter asking him to knock off with the "flowery French expressions," and stick more to "your own simple straight forward & far more agreeable style." Gradually, he did. Almost exactly three years later, Darwin climbed a mountain peak in the high Andes and compared the view to "watching a thunderstorm" or "hearing in full Orchestra a Chorus of the Messiah." That's not just better and more mature writing, it's a sign that Darwin was starting to realize that actually, you *can* comprehend the glorious pleasure of a mountain peak or a Brazilian forest without experiencing it. The scenery had given him an adrenaline shot, but, he realized, perhaps you've got your own soul-awakener somewhere else, at the art museum, or the beach, or even (why not?) the football stadium. Now, as Jean and I stood there, with the entire world seemingly arrayed below us in a tossed mixture of greenery, rock, cement, sand, and ocean, I felt like I had earned a taste of Darwin's Brazilian chaos of delights.

2: PORT DESIRE

Darwin's Rhea

> *What we had for dinner to day would sound very odd in England. Ostrich dumplings & Armadilloes; the former would never be recognized as a bird but rather as beef.*
>
> —BEAGLE DIARY, SEPTEMBER 17, 1832

ANY GOOD DARWIN ACCOUNT should feature the story of the small South American ostrich known as Darwin's rhea, because it so nicely sums up the spirit of nineteenth century adventure. It is, first of all, a story about the thrill of being the first to discover a new species. It is also a story about the thrill of finding that you are accidentally discovering a new species while eating it.

In December 1833, two years into its journey, the Beagle anchored in the expansive natural harbor at Port Desire on a dry, flat coastal stretch of Argentine Patagonia. After leaving Rio de Janeiro, the Beagle had never returned to Brazil and instead spent two years surveying back and forth, up and down the Patagonian coast, arriving now in what was becoming a familiar scene—a deep blue harbor surrounded by wind-tormented desert plains. The ship's recently hired artist, a friend of Darwin's named Conrad Martens, found himself with little to sketch as the Beagle lingered. "The country is bare and desolate in the extreme," Martens wrote in a letter to his brother. Wood for fires was hard to come by, and worse still, the water was brackish and so full of bacteria that Martens felt the need to purify it "with certain proportions of brandy." Putting down his sketchbook, Martens picked up his shotgun and strolled out most days to exercise the local wildlife—simultaneously killing time and animals. Part of the benefit of his activity was to provide some food for the rest of the crew. The Beagle diet generally consisted of anything caught or found that tasted better than salted beef and pork, and at Port Desire this alternative menu included gulls, shags (cormorant-like birds even less appetizing than gulls), sharks, mussels, limpets, and land crabs.

The other benefit was scientific. Martens could bring his animals back to Darwin for inspection, and if they piqued the naturalist's interest, he would send them back to England for classification. It was on a run-of-the-mill January afternoon meal-seeking excursion that Martens unwittingly handed Darwin one of his greatest finds.

Locals had told Darwin several times of a smaller version of the South American ostrich, or rhea, which supposedly roamed the plains of southern Patagonia and which Europeans had never seen before. (The northern plains were full of regular ol' "greater rheas.") Darwin recognized that finding this "avestruz petise" would be a major feather in his species-hunter cap. He also knew that the French government had recently dispatched a man named Alcide d'Orbigny to collect animals in South America, and Darwin worried in letters home that this sinister competitor would "get the cream of all the good things" before he did—most especially the small rhea.

Martens knew nothing of this Anglo-Gallic battle for ostrich-discovery supremacy. Upon seeing a small bird while pacing the plains, his first thought was that no one from the Beagle had successfully hunted an ostrich of any kind, and so, meaning to become the first, he tracked it, snuck up on it, and got his shot. The rhea fell and Martens grabbed it, slung it across his shoulder, and headed back to camp to hear his praises sung. Not only had he brought back the first ostrich, he had brought back an undeniably tasty dinner.

Darwin, like the other crew members, was pleased. He had sampled rhea before and compared the flavor and texture to beef, and after a cursory examination of Martens' bird he absentmindedly concluded it was your typical large rhea, only a juvenile variety. It was "skinned and cooked before my memory returned," he later wrote. When he realized his error he managed to gather "the head, neck, legs, wings, many of the larger feathers, and a large part of the skin," and shipped these back to England, where naturalist John Gould confirmed that these bits and pieces constituted a new species and gave it the name *Rhea darwinii*. "M.A. d'Orbigny," Darwin observed dryly when he got home, "made great exertions to procure this bird, but never had the good fortune to succeed."

In this small matter, though, Darwin was wrong. D'Orbigny (who Darwin hailed in *The Voyage of the Beagle* as an "indefatigable" collector) had, in fact, not only apparently discovered the smaller rhea—he had already named it the *Rhea pennata*, which is the scientific name used today. But by virtue of Darwin's greater status (or possibly just to avert another hundred years war), the bird is still commonly known as the Darwin's rhea.

The Darwin's rhea story was the first story that made me see how much humor, irony, and brilliant accident there was in *The Voyage of the Beagle*. Imagining Darwin's "transport" as he rushed about gathering dismembered bird parts in order to keep them out of the cooking fire made me laugh the first time I read it, at a hostel dining room table in Chile, and again a few weeks after leaving Rio de Janeiro, as I rode on a bus toward Port Desire, looking out the window at the rheas that flocked alongside the highway. I'm not expert enough at birdwatching to instantly spot the difference between similar birds—for one thing, "smaller" is a pretty relative descriptor to go by unless you spend a lot more time with rheas than I have.

So I craned my neck to follow the birds for as long as I could from the bus window, watching them whip past at 50 miles an hour, thinking, was that one shorter? Were its legs more feathered? What, exactly, do ornithologists mean by "mottled"?

I also tried to think, off the top of my head, of another species named after the Europeans who first ate them, which turned out to be a short list. I stopped looking out the window for a while and looked around at the interior of the bus. The rest of the passengers clearly didn't share my enthusiasm for the small, dusty-brown bird's history. Most were shrimpers. They were silent and burly and presumably interested in, well, shrimping. The commercial fishing season had just started, and fish and fishing had occupied the prime real estate above the fold of the Patagonian newspapers for the last week. The land being as desolate as it was, it seemed natural that most of Patagonia's residents focused instead on what was in the water.

When we arrived in Port Desire, and the shrimpers had trundled off to the area around the wharves, I found that the town that inspired Martens to pack away his paintbrushes for a week was now even less picturesque. Despite its location at the edge of a major inlet, there was almost no public waterfront. All the land along the coast had been given over to fish processing plants, navy bases, and shipping containers. Massive freighters rusted into the water, roped together like yoked oxen in front of storage lots full of Maersk and Hamburg Sur containers.

In the late 1700s, the Spanish tried to settle in the area around Port Desire and, as with many of their settlement endeavors in southern Patagonia, failed for lack of water. When Darwin arrived the city was just a ruin on the edge of a big estuary, crumbling and unpopulated, although many of the buildings kept what the naturalist called their "good style." "The fate of all the Spanish establishments on the coast of Patagonia, with the exception of the Rio Negro, has been miserable," Darwin wrote.

Port Desire didn't fare much better for another seventy years, and not until the beginning of the twentieth century did it become established as a fishing town and naval base. The town today has a population of 12,000 but retains a deserted, Old-West ghost town feel. Wandering down the main drag at noon, it seemed quite possible that I was the only tourist for miles around. Not only was it an out-of-the-way and, frankly, largely uninteresting place, but it was the tourist low season, when most of Patagonia sits back to take a nap and admire the tumbleweeds and rheas drifting across the plains. My hotel, one of a handful in town, was seemingly the social center of the city, full of visiting fishermen who took their breakfast around 5 A.M. and were asleep by the time I returned in the evening.

Posters around town advertised an excursion company called Darwin Expeditions, so with nothing else to do, I called to see if they could help me find a Darwin's rhea. I hoped that an expert could not only help me positively identify the smaller rhea, but could put them in context, and tell me more about the birds of Port Desire. But in the low season, Darwin Expeditions only ran boat trips around the harbor. Pressed with my request, the guide, Ricardo Perez, offered a seabird-and-penguin watching cruise, but added that the minimum number of people to do a boat ride was four, and that no one had signed up yet, and that the bus I'd come in on—the one that had been full of non-tourist shrimpers—was the only bus of the day to arrive in town. I hesitated, then decided that even the small chance of a boat ride beat wandering around town studying fish processing plants for another day. Perez seemed to think it unlikely this deafening non-demand would change, but said he'd call my hotel later that evening.

When he did call, around 9 P.M., it was to tell me that four other tourists had come by. Perez sounded about as surprised as I was about this. We were on for penguins.

In the morning, the hotel called me a taxi, which dropped me off at a dock at the edge of town. I stood waiting to see what kind of fellow tourists had arrived and enabled me to go on this boat ride. People who spend a lot of time in obscure corners of the world may be able to anticipate what kind of travelers are around in remote places in the low season and have their own private transportation into town, and those readers may be nodding and starting to smile now at what seemed like the cosmos choosing to have a bit of fun with everyone. Well, I didn't know what to expect. I tried to tick through the list in my mind: no one young, certainly, because they would have arrived on the bus. Europeans or Australians on a guided tour? Wealthy Argentines who had access to a private car and decided on a whim to visit Port Desire? I mean, who (other than Darwin-fan-writers) rents a car in Patagonia? And then comes to a place like this?

Across the highway, a white sedan pulled up and two upper-middle-aged couples got out. They looked like cruise ship passengers: oldish, whitish, dressed in obvious western-traveler duds. They crossed the road and introduced themselves in halting but competent Spanish. Picking up their accent, I asked where they were from, and they turned out (of course!) to be Mormon missionaries from California. They talked for a generous while about their travels and their missionary work, which they were on a break from, but their sons, who were now in Buenos Aires, were still working at it and evidently really enjoying it.

I started to think I was going to be in for a very strange boat ride when I told them what I was doing here.

Which, evidently, was their cue to ask.

"So," said one of the two men, who I'll call Jim. "What brings you to Port Desire?"

This is the thing about missionaries: I'm fairly suspicious about them. Not that I find Mormonism in any way exceptional or offensive, but I just don't much like the idea of evangelical anything. So I worried about what they'd say, and then I thought about it and worried more about what I would say. Introducing the subject of Darwin into a crowd of missionaries seems like an activity fraught with conversational peril. Darwin means evolution, and evolution, in my experience, is not something that you can casually discuss with people who are zealous enough to be missionaries. So I picked my words carefully and said in the most inoffensive way I could think of that I was retracing Charles Darwin's footsteps.

There was a longish pause. I shuffled my feet and

triple-checked the buckles on my lifejacket, and then, to my surprise, they nodded politely and said what a wonderful thing it was that I was getting out in the world and that they hadn't known that Darwin was so young on his trip. We all smiled weakly for a while, and then they changed the subject to birds, at which they were quite expert.

Perez, who had been off fiddling around with ropes and things, came back around this time and helped us up the plank and into the boat, which was some form of g l orified zodiac. Jim, standing in the bow of the boat with a camera, started to point out different varieties of birds to his wife with a really impressive, encyclopedic range. "There's a red-legged cormorant!" he said. "You'll never see one that close!" As everyone else noticed the red-legged cormorants nesting in the pink cliffs, he moved on to regular cormorants and night herons and white-legged somethings-or-others and lots of other birds I didn't recognize. "Finally!" he exclaimed at one point, gesturing at his camera, "a picture of a white-legged-something-or-other!"

His enthusiasm was infectious. I forgot Darwin for a while and grabbed my camera. Penguins nested in crevices halfway down the pink cliffs, and Perez pulled the boat in for a closer look. The rock rose out of the water in an improbable jumble, like a pile of Jello cubes, to a height of about thirty feet. I couldn't figure out how the penguins could have got where they were—fifteen feet above the water, fifteen feet down from the cliff edge. It was like someone had forgotten to remind them that they couldn't fly.

We drifted by the cliffs and then motored over to an island in the middle of the estuary, where Perez estimated there were 25,000 penguins hanging out. Magellanic penguins on land are some of the world's most ridiculous creatures; they waddle, they flap their wings futilely, they cock their heads and look at human intruders with expressions of such confusion and incomprehension that you can't help but burst out laughing. (There's a rumor that if you stand in front of a penguin's path as it's trying to exit the water you can seriously harm it, because it will just *stand* there, stumped, until you move out of its way, by which point it will be suffering from the cold.) Perez beached the boat and the passengers all wandered off in different directions in the *pinguinera*, poking around in the grass, stalking penguins with cameras, and sitting on the beach having tea while birds strutted around us.

When we got back into the boat, I resumed my Darwin focus with an examination of the area geology. Numerous small inlets and dry creeks snaked up from the estuary, lightning-bolt shaped cracks in the pink rock. Beyond the water's edge it was fairly desolate, with acres and acres of wasteland dotted with small thorn bushes, yellow clumps of prickly grass, and a howling wind that whipped across the land and turned the water into froth.

As we motored back toward town, Perez started clandestinely pointing out Darwin sites to me, taking his opportunities while the others gazed the opposite direction at birds. Perez swung around to face the distant head of the inlet, where Darwin and others had camped while exploring a maze of pink-rock cliffs. Closer to town, in a patch of sapphire water in mid-estuary, he idled the motor and identified the spot where the Beagle had anchored.

"The ruins of the fort that Darwin mentioned would have been right about there," Perez said, pointing landward at a flat mound of rock rising behind rusted fishing boats. I watched a group of dockworkers tossing silvery fishing lines off the back of an industrial ferry. The estuary glittered around them under a glaring, cloudless sky. Perez paused for a moment. "Of course," he said, "now it's all gone."

I was disappointed, and it wasn't until quite later that I realized that I'd got it all wrong. I'd felt separate, even smug, about looking for Darwin sites on a boat full of missionaries, and disappointed when those Darwin sites turned out to be mostly gone, and I'd forgotten the point of my trip. Instead of celebrating my evolution-enlightened superiority, I should have talked to the missionaries about what we all—including my long-dead naturalist—had in common: a love of nature, exploration, and travel, and a desire for the thrills that all people find in chasing white-legged-something-or-others. That's not just a way to bridge a gap between a 20-year-old skeptic and 60-year-old evangelical, it's a way for everyone, religious or not, to better understand the real Darwin. Who, if he had been along with us on the boat ride, would have been down at the end of the boat with the missionaries, checking out the birds.

3: PORT SAN JULIAN

The Patagonian Myth

Port San Julian was the dramatic site of the most notable acts of first contact between the white man and the Argentine land, and in its barren beaches was written, with the blood of natives and Europeans, the first pages of the dark prologue of Argentine History.

—PABLO WALKER, PUERTO SAN JULIÁN, ORIGEN DEL MITO PATAGÓNICO

THE BEAGLE'S CAPTAIN, ROBERT FITZROY, had decided in 1833 to hire a second ship to help speed the survey of the Patagonian coastline. After several months of hard work he was getting antsy, and he had come across a sealing boat while visiting the Falkland Islands and decided to purchase it. For the next year, both the Beagle and the new addition, the Adventure, ferried their explorers around, taking measurements and making charts. On one of these small surveying voyages, FitzRoy determined the Adventure was not sailing well enough, and so when the ships arrived at Port Desire, he ordered it taken out of the water to have its sails altered. The repair delays mounted, and as the Adventure languished, the Beagle took off again. FitzRoy and crew would survey the harbor at Port San Julian, 110 miles to the south. Darwin, of course, went along.

Although there was no settlement then at Port San Julian, the harbor had managed to cram a lot of history into its short, unhappy life. It was discovered in 1520 by Ferdinand Magellan and his fleet of five ships. Just north of Port San Julian, Magellan encountered the Tehuelche natives of the region and of them his chronicler, Antonio Pigafetta, wrote, "One of these men, as tall as a giant, came to our captain's ship to satisfy himself and request that the others might come. And this man had a voice like a bull's." Magellan's ship then encountered a ferocious storm, emerged unscathed and sailed into Port San Julian on March 31, 1520, to spend the next five winter months. Pigafetta continued his note-taking. "We saw a giant who was on the shore, quite naked, and who danced, leaped, and sang, and while he sang he threw sand and dust on his head." Magellan sent crew members out to lead this giant back to him, which they did. "And he was so tall," Pigafetta wrote, "that the tallest of us only came up to his waist." Magellan named these giants "Pathagoni," a word that, according to a footnote in the R. A. Skelton translation of

Pigafetta's work, means "dogs with large paws" in various Romance languages. The name for their land became Patagonia.

Almost immediately after arriving in the harbor, the masters of the other four ships mutinied. The mutiny failed, meriting scarcely a mention by Pigafetta, and Magellan beheaded two of the lead mutineers. Then, for good measure, he had them drawn and quartered. He hung their remains on wooden crosses as a warning to others, smack in the middle of an island in the harbor, which became known as "Magellan's Gibbet" (a gibbet being a useful little device, often cross-shaped, for hanging or displaying remains). He named the island *Isla de la Justicia*, the Island of Justice.

When the long winter came to an end, Magellan trekked to the top of the highest mountain in sight, planted a cross on top, and claimed the land for the King of Spain. He sailed on to "discover" the Pacific Ocean, give it a name, and then get hacked to death by an angry mob in the Philippines. (A sadly recurring theme for the great explorers; Pigafetta's description of Magellan's death—a confusing melee in which the captain fell on the beach as he was repeatedly stabbed—is a tableau familiar to fans of Captain Cook.)

Sixty years later, in 1578, the English pirate Sir Francis Drake sailed into the harbor at San Julian. Drake, too, had a suspected mutineer on board, so he called together a jury of forty sailors from his various ships, and the jury found the man guilty. Drake had him beheaded and hung, also on the Island of Justice, and etched a few words commemorating the deed into a rock on the island.

The bay continued to attract famous visitors: English circumnavigator George Anson came in 1741 and made it his base of operations in case of an attack by the Spanish. Spanish city-founder Antonio de Viedma, who helped colonize much of Argentina, tried (and failed) to establish a city called Floridablanca here in the 1780s.

Darwin arrived on January 9, 1834. For a week, the Beagle had explored the coast, and while sounding the sandbars in the harbor, FitzRoy let Darwin off the ship to examine the geology. A few days later, he went out walking and made two discoveries he could not easily explain: "a Spanish oven built of bricks, & and on the top of a hill a small wooden cross." "Of what old navigators these are the relics it is hard to say," he wrote in his journal. Although he knew both Magellan and Drake had visited the bay, he did not mention Floridablanca, which seems the most likely origin of the Spanish oven. The cross could have been Magellan's—if the wood had survived for 314 years—although Darwin noted that it was small. Pigafetta had observed that the cross Magellan placed on the hilltop was "very tall." (And a Spaniard would never lie about the size of his cross.)

The bus terminal at modern Port San Julian was an afterthought and, for most, a fifteen-minute bathroom break on the long haul between Buenos Aires and Rio Gallegos, the southernmost city in mainland Argentina. Going north, the bus pulled up at Port San Julian at 2 A.M.

and left twenty minutes later. Going south, the bus arrived at 1 A.M. and left after a mere ten minutes.

Darwin called the San Julian harbor "fine," which, in the tourist literature, local historians had translated to "beautiful." In the dark the town looked like most other Patagonian towns, all low buildings made of cement and corrugated metal, lit with glowing orange street lamps. I arrived at 2 A.M., one of the more lively times of day, since that's when everyone was arriving or leaving. The coffee shop in the bus terminal was open and packed, the different bus companies had staffed ticket booths, and cars cruised the streets. Exhausted by a sleepless bus ride, and still not operating on Port San Julian time (i.e. being awake and alert at two in the morning), I lugged my pack down the street to what looked like a hotel (it had a glass front door and a bar), inquired and received a room, locked myself in, and promptly fell asleep for the next ten hours.

When I woke up around noon the next day, the stores were closed, shuttered, and locked, the sidewalks were empty, and the cars and their drivers had deserted the streets. It was Sunday, always a quiet day in heavily Catholic Latin America, but I walked for several blocks without passing a person or open business, arriving after ten minutes at the harbor. A freshly painted replica of Magellan's flagship, Victoria, creaked and groaned in the wind. Wax sailors hung from the rigging. The harbor extended around the town, which was on a peninsula, and I pulled out a copy of one of Martens' drawings, labeled "entrance to the harbor at Port San Julian," and compared the view. The low-profile town did almost nothing to change the scenery overall, ending, as it did abruptly, in the low brown hills which still looked very much as rendered by the Beagle's artist. The tallest hill, where Magellan put his cross, was visible at the right edge of the drawing. Now called Monte Wood, the summit's distinctively flat top helped me recognize it instantly, even before I saw the large metal cross poking up into the blue sky. With the entire town closed for Sunday, I decided to do what any nineteenth century naturalist would have, and walk to the top of the distant hill.

One thing I envied about Darwin: Randomly walking places must have seemed quite natural to him. As a product of twentieth century U.S. suburbs, I found it strange to glimpse a mountain, set my sights on it, and just start walking toward it on the most direct, as the crow flies, route. I kept thinking there must be an officially sanctioned trail, or a road somewhere, because that's all I had known. North American national parks are so crisscrossed by trails it seems no natural wonder exists without a well-marked trail leading right to it. There's no real need to think about routes, or even to watch your step. But in South America, there are plenty of places—like Monte Wood—where the only way to get there is to blaze your own trail. You learn to appreciate distance that way. You start weighing your desire to experience distinctive parts of the landscape against how long it would take you to walk to them. Months later, back at home, I would look at the hills on the San Francisco Bay peninsula and think, "I could probably be up that and back by lunch."

The sun burned down from the clear sky, and it would have been scorching were it not for the hellacious headwind hampering my progress. I passed the town sewage treatment plant just outside of town, and the dump, where a trash fire was blazing. A cloud of putrid crap-and-refuse smoke was pushed by the winds back across the city.

After the dump, the road devolved into a standard Patagonian gravel rut. The air was so dry I could feel my lips cracking after about twenty minutes and the dust from the road settling in my mouth and around my teeth.

On January 11, 1834, Darwin and FitzRoy and crew set off into the same terrain in a search for fresh water, using an old Spanish map. They walked all day but couldn't find a drop, and FitzRoy became dangerously fatigued. Darwin, who was more accustomed to long hikes, lit out for a lake a few miles in the distance but found to his "great mortification" that the lake was nothing more than "a field of solid snow-white salt." Upon receiving this news, the crew decided they could do nothing else but return to the ships. FitzRoy, however, could not make it. "About dusk I could move no farther, having foolishly carried a heavy double-barrelled gun all day besides instruments," he wrote. "So, choosing a place which could be found again, I sent the party on and lay down to sleep; one man, the most tired next to myself, staying with me. A glass of water would have made me quite fresh, but it was not to be had."

Darwin struggled back to the Beagle with the others and sent back help for FitzRoy and friend. "Towards morning we all got on board," FitzRoy continued later, "And no one suffered afterwards from the over-fatigue, except Mr. Darwin, who had had no rest during the whole of that thirsty day—now a matter of amusement, but at the time a very serious affair." Darwin recorded a slightly different version of the day's events—"I was not much tired," he wrote in his diary—but he quickly contracted a fever and was consigned to bed for the next two days.

In analyzing the difficulties of that abortive walk, Darwin concluded that the climate had done them in: "as we were only eleven hours without water, I am convinced it must be from the extreme dryness of the atmosphere."

Someone had erected a fence about twenty yards below the summit, but since there was no one to object, I crawled under the barbed wire and quickly hiked the rest of the way up to the wind-blasted peak. With nothing to shelter behind, I worried that my fifteen-pound backpack would blow away and balanced by hunching at a slight angle into the wind.

The giant metal cross I had glimpsed from the harbor below was made of a steel lattice, and the wind whistling through it made a sound like a subway train passing through a tunnel. I could see all the way down to the town and across the bay to what was purported to be Magellan's Gibbet and Island of Justice.

I wondered if Darwin had climbed the same hill. It seemed to me almost inconceivable that he would not have, although he never mentioned it by name in his journal (the name Monte Wood was already being used by that time, which we know because FitzRoy mentioned it as the landmark he used for finding the harbor). Darwin did mention climbing two hills while in Port San Julian, the first when he found the wooden cross, and the second where he and FitzRoy first glimpsed the salt flats. Either could have been Monte Wood. Darwin might not have recorded the name in his journal simply because the hill, I had to admit, was not particularly impressive.

The next day, Monday, was a holiday, and I wandered aimlessly around town for a while, hoping mostly that a grocery store would open and allow me to buy my favorite budget traveling meal, bread and cheese. But with the doors still locked and barred at 11 A.M. I gave up and moved on, making a note to think about trying again when the out-of-town buses arrived after midnight. Next, I tried the city history museum, and it was locked, and then I tried the archaeological museum, and that was locked. I tried the tourist information office, where a woman was sweeping the sidewalk in front—it was locked. She turned out to be an employee of the hotel next door, although her hotel was closed for the time being. I wandered along the bay and returned to the replica Victoria, whose ticket office was boarded-up, and then came to the only guided tour company in town. Reminiscent of Darwin Expeditions in Port Desire, they advertised dolphin-watching expeditions, but they too were boarded-up and locked. A faded dry erase board on their storefront heralded “Next departure:”—followed by a large blank spot.

I wandered back up the main street, San Martín, which runs from the highway down to the bay. It is really the only street in town—the roads are paved for about two blocks on each side of it and beyond that it’s dirt and gravel and low cement houses. In the middle of the block I found another tourist information office, and although I didn’t see anyone, the door was open.

Inside, there was an empty desk with a computer and a letter-sized printed report with the title: “What are we doing about tourism?” The cover had a picture of Rodin’s “Thinker.” After walking around in the wind, it was startlingly quiet in the office. The large windows had a view up the mostly empty street.

In the other room of the office, I found a woman sitting at a desk.

“Hi,” I said. “I’m looking for something to do.”

She smiled and nodded knowingly. She looked tired. She handed me a brochure that I’d already been given by the receptionist at the *hostería municipal* (my hotel, the descriptively named, “municipal hotel”), showing sites of interest that included all four restaurants in town (all four of which were closed), the bus terminal, the tourist information office, the police, and a few hotels (most of which were also closed).

“I noticed that the museum was closed,” I said.

“Yes. The museum is closed.”

“Will it open?”

“No. It’s closed.”

I pondered this for a minute. There went half my plan for the day. "It seems," I ventured slowly, "like there aren't many tourists around."

She shook her head sadly. "No." It sounded very tragic, the way she said it.

"So," I said. "What do people do here for work?"

"Most of them work in administration," she said.

That made sense. The only buildings I'd seen open were social security offices and banks. But it still didn't help me plan the rest of my day.

"So," I said, returning to my theme. "What can I do?"

"You could go to the archaeological museum," she said hopefully. She gave me another brochure and pointed it out on the map. "You could go to the nautical museum, but that's also closed. But you could take pictures, or look at it from the outside."

I didn't mention that I'd already tried both places. I told her that I was researching Darwin. Was there anyone I might talk to?

She thought for a minute. "Pablo Walker," she said. "You can find him at the university."

I went to the university to look for Pablo Walker.

The university was about the size of my elementary school, one story tall and painted a kind of pale mint green. I asked after Pablo Walker at the reception desk.

The receptionist eyed me suspiciously and asked me to wait while she checked. She disappeared through the cafeteria and returned a minute later. "Follow me," she said.

She led me through the empty cafeteria—the kind of multipurpose, vinyl-floored auditorium we had at my junior high school—past a small room marked "Professors" on the door and consisting of a single desk, bookshelf, and two computers. We walked through another short hallway and entered a classroom that had about thirty small desks pushed together into the corner. A larger desk in the center of the room supported two computers wired to some video editing equipment. Three men sat around, staring at monitors. The receptionist waved me in and went back to her desk.

One of the men stood up when I walked in. He was wearing a black sweater vest over a white T-shirt and faded, dusty, cuffed blue jeans. His hair was gray and brown and parted into a wave that soared off the top of his wrinkled forehead. He was wearing thick metal glasses and he had a thick red beard that didn't quite cover his chin.

"I'm Pablo Walker," he said, shaking my hand. "Can I help you?"

I told him that the woman in the tourist office had provided his name and that I was interested in Darwin and the history of Port San Julian.

"Sure," he said. "I'm working until 5:30, but come back then and we can talk."

Pablo Walker was still sitting at the video-editing desk when I returned at 5:30. He came over to the desk where I had parked myself and sat down opposite me.

Unsure where to begin, I asked what kind of professor he was. "History?" I guessed.

"Actually, I'm not a professor. I'm just a lecturer," he said. "I just read a lot. I'm a fan of history."

"Walker isn't exactly a common name in Argentina," I ventured.

"Well, my great-grandparents were some of the first immigrants to come to San Julian, and they had come originally to Punta Arenas from England. I'm the fourth generation to live in San Julian, and my great-grandparents were among the first thirty people here."

"Are people here interested in Darwin?"

"People are interested because of tourism," he said. "But you look at the bigger names, like our street names. Like in the U.S. you have Washington, our streets have names like San Martín. We have a Darwin Street, but it's a secondary street."

Walker remarked that until three years ago, he had been the tourist director for San Julian. During that time, he had overseen the building of the replica Victoria. It was a hard-fought battle, he said, to keep other tourism officials from turning the Victoria into a Disney attraction, more cartoonish and less historically accurate.

"Now I have a project to build a replica of the Beagle, for a museum," he said. "Like the Victoria, except that's more nautical. This would be more of a science museum."

"What do people in Argentina think of Darwin?" I asked.

"In Argentina, they accuse Darwin of believing Patagonia to be bad land," he said. "Darwin saw that there was not much life in the interior, but he didn't see how much life was in the water. He was focused on geology. The coast of San Julian has 75,000 penguins. There are seals, cormorants, dolphins. It's a bit strange in the case of San Julian that Darwin focused on geology and the land."

"And do you study him in school?"

"My classes, or in general? In general, we study him, because wherever you are, he's on the list of ten most influential scientists. In my classes, I teach a bit of the history, and take people to where he was, so they can see, take pictures and understand things that he saw."

I asked how much time he spent on Darwin in his classes. He said that a course met twenty times during a typical semester, devoting five sessions each to "the four most important things here": Magellan, Drake, Darwin, and Floridablanca.

"And which of those is your favorite subject?"

"Darwin."

"It's interesting," I said, "that there's so much history in this one small town."

He nodded. "The coast of Santa Cruz is very inhospitable," he said. "There are only five or six places where they have safe ports: Puerto Deseado, San Julian, Puerto Santa Cruz, Rio Gallegos, and Rio Coyle." He ticked each port off on a finger.

"Of those, San Julian has the best port. It's small, but for small boats like the Beagle, or Drake, it was perfect. That's why the history is so concentrated in such a small area. What's lacking in San Julian is the history before Europeans," he continued. "There were people here 8,000 to 11,000 years ago. Patagonia has had people in it for longer than Brazil. It's sad that for the original inhabitants there is nothing to study."

There seemed little hope, then, in learning more about Magellan's *Pathagoni*. Since I had another full day in Port San Julian, I asked Walker what he would recommend I do on Tuesday. "Come back around noon," he said. "We'll go out and see the places where Darwin went."

The next morning dawned clear and bright, with a rampaging fifty-mile-an-hour wind that made being outside nearly intolerable. I met Walker at the university and we drove to his house in a beaten, faded 1970s-vintage Volkswagen sedan. There, he switched to a new white van, and we set out along the coastal road toward the mouth of the harbor.

Outside the wind was making things miserable, but from inside the van, it had the effect of clearing the air and opening the view. We could see across the harbor to the cliffs on the other side, where Darwin had "geologized." Walker pointed out the different sedimentary layers and told me the age of each.

"The point is called Punta Asconapé," he said. "Point Shingle." His English was good—he spoke it fluently—but for whatever reason, he chose not to talk to me in English unless it was directly relevant to his lesson.

"I think this point was named by FitzRoy after talking to Darwin," he said. "Darwin wrote while he was here about the accumulation of pebbles from the Andes on the plains, called shingle. FitzRoy named that Point Shingle, for Darwin."

Walker was unafraid to offer his own opinion about the city's history. I had gathered from reading newspaper stories about the Victoria replica that this did not always make him popular—he was inevitably labeled "the revisionist historian Pablo Walker"—although when I asked him about this, he shrugged and said it was mostly politics.

We drove around the base of Monte Wood. I told him I'd climbed it on Sunday and he said he also had climbed it, a few hours after I had. "Windy up there," I said.

"And it was calm in town on Sunday," he agreed. "Today, you wouldn't be able to stand."

"It makes sense that Darwin climbed it," he added, without prompting. "It was the highest point. If he wanted a view, he would have climbed it."

He slowed down and pulled to the side of the road. "See the penguins?" he said, pointing at an island in the bay. Small groups of six or seven penguins stood around, wings slightly raised in a gesture that—and here I could be anthropomorphizing just a little bit—indicated confusion.

"In the right season, there are 75,000 penguins on that island," Walker said. "Right now, the only ones who are still there are the ones with problems. Maybe they're sick. But if they're still there in a few days, it's very likely they'll die."

The island appeared on maps as Cormorant Island. Walker suggested that it was the real Island of Justice—and that the place that bore the label of Isla de la Justicia was not. He suggested that the English captain George Anson mixed up the map of the harbor, confused some of his compass directions, and so reversed the islands.

"When the tide is high, Isla de la Justicia is very small," he said, as evidence. "It's only about fifty meters across. There's almost nothing there. But look at this island, it's much larger."

"But can't you find Drake's rock on Isla de la Justicia?" I asked, referring to the rock on which Drake had carved a Latin inscription to commemorate the execution of the mutinous sailor.

"If that actually was the real Isla de la Justicia, which I don't believe it is, yes, you might," he said. "But no one has. And it's been five hundred years now." Walker said he remembered Bruce Chatwin, the author of the famous travelogue *In Patagonia*, visiting San Julian in the 1970s. Chatwin devoted a paragraph to his own unsuccessful search for Drake's rock. (Chatwin, not a Darwin aficionado, wrote succinctly, "I passed through three boring towns, San Julian, Santa Cruz, and Rio Gallegos.")

Leaving historical mysteries aside for the moment, Walker again stopped the van at a small rocky point called Cape Curious. We got out and walked along the rocks, which were overflowing with fossil mussels and small shells. "These rocks are about 40 million years old," he said, pointing out different layers in the cliff. We walked along the cliff, peering at different kinds of fossils, for a few minutes. The wind picked up speed and blew the tops off the breakers, sending a plume of spray forty feet into the ocean. Sand got in my eyes even though I was wearing sunglasses. I felt it stick to my ears and neck.

When we got back in the car, I grunted. "The wind is really blowing."

"Probably sixty or seventy kilometers per hour," Walker said. "It can reach up to one hundred."

The ocean was heaving and looked remarkably like the picture that the Beagle's ship's artist, Conrad Martens, had drawn of the harbor. The idea of being out on a small boat in that wind didn't appeal to me at all, and I can imagine Darwin's sensitive stomach finding it quite disagreeable. Darwin didn't mention seasickness, although his entry from January 15 reads: "A heavy gale of wind from the SW; several breezes from that quarter have reminded us of the neighborhead of Tierra del Fuego."

Walker and I continued driving around the cape to an Atlantic beach, where the remains of British carbon mines still stood in the cliffs. Walker pointed out carbon on the beach; long stretches of black rock that broke off easily. "It's not very good quality," he said. "But the rocks here are very metallic. Feel how heavy this is." He picked up a rock and handed it to me, and my arm sagged under the weight. When I dropped it, it made a metallic plink on the other rocks.

We walked along, looking at more fossils. "This is the oldest formation here," Walker said, pointing out to the end of the beach. "Seventy-five million years old."

I could see Darwin walking the same beach, delighted by the profusion of fossils. We saw football-sized mussels, long-since extinct. We saw sand dollars—"these are also forty million years old," Walker said, "but you still have live ones in North America, right?"

The cliffs were lined with fossils up to heights of forty or fifty feet. We walked out to the point, Walker leaping between rocks and striding purposefully across the beach, each time with a specific thing in mind to point out. He had obviously given this lecture a few times before.

Finally, we came to the end and looked out at the raging ocean and up at the fossil cliffs.

"Sadly," he said, "that's all the time I've got today. Back to work."

He drove me back to the university and there he gave me a short historical paper he had written. With nowhere to go, and relieved to have some human company in an otherwise lonely town, I stuck around and started to read his history, while Walker and another professor hunched over a plastic model airplane, which they had positioned in front of a beige sheet under a set of bright camera lights. They had a small camcorder on a tripod pointed at the airplane, and they turned this on and filmed for a while. Then they moved the camera over to an editing station where Walker cut the plane out and inserted it into a video shot earlier at the Port San Julian airport. The second professor pulled out a thermos and started passing around tea while they settled in to watch the finished product on a pull-down screen. The screen went dark, then flashed the title, "Pilotas de las Malvinas." And then dramatic action music started, a mix that sounded suspiciously like the theme to Hollywood action blockbuster *The Rock*. A quick zoom caught a heroic pilot walking across the tarmac in his jumpsuit. Another zoom captured his manly jaw and hard, squinting black eyes. He looked determined as he walked past a set of warlike planes parked in the hangar—a live action male striding purposefully past the plastic model that Walker had inserted only a few minutes ago. I looked at the raised canopy on the parked planes and then over at the raised fake-glass canopy on the model at the other end of the room.

This was weird, the kind of thing I used to do when I was twelve years old, not the kind of thing I had expected in an academic setting. But there was a huge qualifier: This was a film about the Falkland Islands War. And when the Falklands are involved, in Argentina at least, normal feeling doesn't really enter into it.

A brief historical diversion concerning the small island group off the coast of southern Argentina: On March 1, 1833, the Beagle arrived at Port Louis, in the Falkland Islands, after a long journey through Tierra del Fuego. "The first news we received was to our astonishment, that England had taken possession of the Falkland Islands & that the Flag was now flying," Darwin wrote. He offered his own brief history of said event: that the islands had been uninhabited until the Argentine government (which Darwin referred to as the "Buenos Ayres Government") sent settlers there, which in turn prompted England to respond and reassert its claim to the islands, which they had stumbled upon and claimed as their own—along with Spain and France—in the 1500s.

When the Argentine governor of the Falklands felt forced to surrender them to the English in 1833, he had little power to resist, which didn't stop him and other members of the government from grumbling loudly. "By the aweful language of Buenos Ayres one would suppose this great republic meant to declare war against England!" Darwin scoffed. He little understood the depth of feeling Argentines held for the islands, a pride that surfaced again in 1982, when the military dictatorship of Argentina decided to make another grab for them. Unfortunately they underestimated the depth of feeling English Prime Minister Margaret Thatcher nurtured for the islands too. From my view, both were hard to understand. (The famous Argentine writer Jorge Luis Borges once remarked that "the Falklands thing was a fight between two bald men over a comb.") What was supposed to be a quick and painless invasion turned into three months of war, handily won by the British, even though they were fighting 8,000 miles from home. Argentina suffered nearly three times as many casualties as the British—635 to 255—and their World War II-era military equipment couldn't match England's high-tech arsenal. Argentina's military government resurrendered the Falklands in June 1982 and their authority quickly collapsed, leading to a new civilian rule in Argentina.

Twenty-five years later, the Falklands still fascinate the Argentines, and their passion is astonishing. Ubiquitous graffiti demands that the British ship off once again, and towns are cluttered with memorials and plaques pledging "We will return." Argentine maps invariably display the name "Islas Malvinas (Arg.)" near the Falklands—the same treatment, incidentally, granted the large sector of Antarctica that Argentina claims, as well as the South Georgia Islands, the South Orca Islands, and the South

Shetland Islands. An English friend who had been to Argentina told me that when he reported a stolen wallet in the town of Mendoza, the police shrugged and informed him, "We'll give you back your wallet when you give us back the Malvinas."

When the film ended, I turned my attention back to Walker's history paper. He had titled it "Port San Julian, origin of the Patagonian myth," and in it he made a fairly compelling argument that Darwin had been one in a long line of explorers to give outsiders the wrong impression of Patagonia as a harsh, sterile, wild land. The first subhead read, "Magellan lands in San Julian, and the cursed legend is born." Magellan's bloody mutiny and red-painted cannibal giants "made an impact in our collective imagination," Walker wrote, and the explorer's fantastical account, combined with the only other events outsiders ever seemed to know about Patagonia, including Drake's harsh justice, failed settlements up and down the coast, and Darwin's descriptions of plains "pronounced by all wretched and useless," only fortified that myth.

"They can be described only by negative characters; without habitations, without water, without trees, without mountains, they support merely a few dwarf plants," Darwin wrote. But the myth had gripped him too. "Why, then, and the case is not peculiar to myself, have these arid wastes taken so firm a hold on my memory?"

It was hard, though, when I later walked down the deserted Darwin Street, with the wind popping in my ears and the dust in my eyes, and all the stores locked and shuttered, and all the houses made of corrugated metal to withstand the elements, not to look around and think, *myth*?

Walker's answer was to look to the water. But although Walker built his defense of Patagonia around the harbor at Port San Julian, it wasn't the focus of Darwin's Patagonia experience. The young naturalist spent only a few days there, and his remarks to his diary were nothing out of the ordinary. The accusations that Darwin found the land cursed came instead from his account of one of his greatest overland adventures, a trip up the River Santa Cruz into the very heart of the Patagonian legend. That river is still largely unvisited, and its mouth on the Atlantic is only about seventy-five miles south of Port San Julian. But just you try getting there.

4: RIVER SANTA CRUZ

The Terribly Uninteresting Land

Already is the change of weather perceptible. Every one has put on cloth cloathes & preparing for still greater extremes our beards are all sprouting. My face at present looks of about the same tint as a half washed chimney sweeper. With my pistols in my belt & geological hammer in hand, shall I not look like a grand barbarian?

—FROM A LETTER TO SUSAN DARWIN, JULY 1832

FEW TRAVELERS ARRIVE IN RIO GALLEGOS intentionally. The last mainland stop before the Strait of Magellan to the south, and the capital of Argentina's far-southern Santa Cruz province, Rio Gallegos huddles around one main street, named after the minister of war who carried out a ruthless extermination campaign against the Indians. (His name and title, General Julio Roca, makes him the intimidating-sounding "General Rock.") It receives a steady flow of visitors who stop there only to leave again, usually a few hours later. In my small residential hotel, instead of the standard introductory "Where are you from?" exchange, waylaid guests greeted one another with: "And where are you going tomorrow?"

Normally, the answer referenced one of three places: Glacier National Park to the east, Tierra del Fuego to the south, or Buenos Aires to the north. From the look of the woman in the Avis rental car agency, the road along the River Santa Cruz was a novel choice. She stared at me like I'd just told her I wanted to drive from Tulsa to Omaha for the scenery.

"Why?" she asked. "There's nothing there!"

This is why. A few months after leaving Port San Julian—months filled, as usual, with more coastal surveying—Darwin and a crew of sailors from the Beagle journeyed up the fast-flowing river cutting through a roughly 150-mile swath of Patagonia. The river wasn't then and isn't now an obvious destination. The previous Englishman to try exploring the river had made it only 30 miles before running out of food. "Even the existence of this large river was hardly known," Darwin wrote, and it's still absent in guidebooks—or at least, in the *Lonely Planet* and *Rough Guide* I carried with me. As I read the footnotes in my edition of the Beagle diary, I found that Darwin and his shipmates had gone further up the river than any previous European explorers, and had nearly discovered its source, a huge and now touristy lake in the glacier-lined mountains near the Argentine border with Chile. Since then, while the lake had boomed, adding an airport and a major highway that connected it back to Rio Gallegos to the southeast, the area along the river had been sectioned off into expansive, open sheep ranches. Based on my map, the finest I had been able to find at the nearby gas station, the road running along the river's edge seemed in places to be entirely theoretical and was marked mostly as unimproved dirt.

Before he actually arrived there, Darwin thought this great open spaces thing was, well, great. His preconceptions of Patagonia as a wild, unknown land fired his imagination, and his letters home scarcely concealed his excitement. "I long to put my foot where man has never trod before," he wrote to his sister Catherine, "and am most impatient to leave civilized ports." In a similar note to his sister Susan, he wrote that FitzRoy had proposed a "glorious scheme" to journey up a previously unexplored river. "I cannot imagine anything more interesting," he concluded.

In August 1832, Darwin's correspondence halted abruptly as the Beagle explored the Atlantic coastline south of Buenos Aires for the first time. Two months later, Darwin the correspondent reemerged from the wild with new knowledge and a slightly different take. "I had hoped for the credit of dame Nature, no such country as [Patagonia] existed," he wrote in a terse note to a professor friend. "In sad reality we coasted along 240 miles of sand hillocks; I never knew before, what a horrid ugly object a sand hillock is."

Darwin had stabbed right at the heart of the Patagonia conundrum: When romantic dreams lead you out to wild open spaces, you soon realize that they're wide open spaces. Often vast and desolate. The scenery gets monotonous in a hurry. And by April 1834, as they approached the mouth of the River Santa Cruz, Darwin's initial excitement had been tamped down by a year-and-a-half's worth of walks on the "terribly uninteresting" Patagonian plains. Halfway up the river, he wrote in his journal, "The great similarity in productions is a very striking feature in all Patagonia. The level plains of arid shingle support the same stunted & dwarf plants; in the valleys the same thorn-bearing bushes grow, & everywhere we see the same birds & insects. . . . The curse of sterility is on the land."

Inaccurate as natural history-minded Argentines like Pablo Walker found this, the woman in the car rental agency didn't seem to disagree. She tried to talk me into a nice plane ride to El Calafate, the lovely tourist destination on the lake where foreigners could speak English and be understood, and see big glaciers, too. I explained to her that I was trying to follow Darwin's path up the Santa Cruz and that to do so I would need to visit Port Santa Cruz, the town at the river mouth, and then drive from there along the river until I got to Lake Argentina, just past where the Beagle crew turned back. She tried again.

"Wouldn't you prefer to go to Piedrabuena?" she asked. "It's a bigger town. It's very nice. And the national park, Monte Leon, is precious."

I persisted.

"At least," she said, "let me call and make sure that road still exists."

Half an hour later, I was in the car, a silver Chevrolet Corsa notable as one of the American cars that you cannot purchase in America, most likely because Americans in big trucks would accidentally drive right over them without noticing. I turned the ignition and puttered out of the rental car lot while the woman waved after me and called "Good luck!"

Ten miles into the drive, I turned to counting dead armadillos for amusement. I started telling myself jokes—dead armadillo jokes. ("Why did the armadillo cross the road? It didn't—it got hit by a truck before it got there.") The horizon—stretched across a dry, flat, endless expanse—flickered, like a mirage. The road ahead met the sky to create another mirage effect, so the roofs of approaching cars came into view several seconds before their lower halves did. Rheas picked along the fences at the highway's edge, and grazing guanacos darkened the horizon. Sheep and birds gathered around trickles of mud, reveling in the moisture.

My arrival at Port Santa Cruz six hours later did little to dispel my feelings of adventuring off into the middle of nowhere. The town had a population of about 4,000 on the southern edge of a fairly wide delta. It looked like an unfinished housing tract—manicured grass and fountains enhanced the median strip of the main street, but the houses emptied into dirt lots strewn with trash. The buildings on the outskirts of town were often half-completed, exposed brick and mortar with wood-shingle roofs. I drove straight to the edge of the estuary. The ebb tide exposed a huge stretch of black sand with rivulets of water draining back into the river.

A rough paved road ran parallel to the water's edge and into the town's municipal tourist office. When it comes to places to stop first in small Patagonian towns, travelers are a bit starved for choice. I tended to end up in tourist information offices, just as a way to confirm that, yes, there was actually a town here, and yes, it was the same town I had been trying to get to. Which the covey of friendly, helpful women in Port Santa Cruz were eager to do. I asked about Darwin, and they agreed, excitedly, that he had been on the river. Definitely. They offered to show me around the one-room history museum next door, and I agreed, then felt forced to feign enthusiasm for stuffed shorebirds and displays detailing notable sheep-shearing exhibitions of years past. I checked my watch. Since it was getting late, and I thought the river mouth would make for a good place to begin the next day, I asked about staying for the night.

Yes, they assured me, they had hotels. They asked if I would like them to call the hotels for me. I said sure. One woman picked up the phone and called four different places, all of which were full or closed. She slumped back in her chair, causing the hinges to squeak. "That's every hotel in the city," she said. "All full."

She fidgeted with some papers on her desk and then said I'd probably be able to find something in the only other nearby town, the oddly named Cmte. Luis Piedrabuena, on the north bank of the river. I thanked her and went back out to the car, hopped in, and started driving again. The highway, shining with reflected late afternoon light, crossed over the river on a long, high bridge, then a short series of side roads curled back to Piedrabuena, just upstream from the delta.

The town looked about the same as Port Santa Cruz. Same brick-and-metal housing, same manicured median strips. This time, the medians had pine trees and bathing nymph marble garden statues. As the sun started to set, I walked down to the poplar-lined river's edge. Two things noted by Darwin—the speed of the current and the color of the water—appeared to be accurate still. The color was a milky, minty green, sediment-rich and turbid. In the river's middle, it flowed around the same thicket-covered sandy islands that had caused the Beagle crew a headache.

"We tracked but a short distance," Darwin wrote on the second day of the trip, "for there are in this part many islands, which are covered with thorny bushes, & the channels between them are shallow, these two causes hindered us much." I peered into the fading evening light and thought, "Check."

Another major nuisance for the Beagle crew manifested itself to me in the *alta peligro* (high danger) sign on the riverbank warning against swimming. The river's swift current meant the Beagle sailors were not able to row their whaleboats up the river and had to pull them instead, the entire crew taking turns in the yoke. Dragging against the river they managed only about ten miles a day, according to Darwin, who made sure to emphasize that "*every one*" (his italics) took a turn pulling. The excruciatingly slow pace ensured plenty of time for studying the repetitive scenery.

I spent that night in a completely deserted hotel, a low, cement building in which, after I received my room key, I never again saw the receptionist. I woke up when it was light out and found a hotel employee smoking a cigarette and reviewing account books in the lobby. He nodded a curt goodbye as I carried my pack out to the car.

I had to backtrack a bit to the southern edge of the river to get to the turnoff for Provincial Route 9, the road that started about fifteen miles inland and extended west all the way to Lake Argentina and the tourist town, El Calafate. I had been making armadillo jokes and feeling end-of-the-earthish on the main paved highway out of Rio Gallegos, but now, as I turned onto the washboard dirt route, I felt like I was leaving the world entirely behind. The sky turned dark. Only a narrow band of blue at the horizon slit the heavy gray above. The clouds had the effect of dimming the lights across the plains, turning green bushes black, so it looked as if a fire had consumed the entire area. Guanacos, sheep, and rheas milled about alongside the road, clearly not used to cars. I had to honk them off the road in places, and many of the rheas would run along the road just in front of me before peeling away at the last minute. I had enough time to pull out my camera and, without needing to worry about oncoming traffic, take pictures of the birds running away from me on the road. Driving behind another, and trying to maintain an even distance behind it, I clocked it on the Corsa's speedometer at nearly twenty miles per hour.

The guanacos, similarly unafraid, stood on the horizon in massive herds, ears perked at the noise of the approaching car, and ran off the road in groups, pulling their skinny legs together in unison and extending their long necks forward like giraffes.

"The Guanaco is in his proper district," Darwin noted near the beginning of the Santa Cruz trip. "The country swarms with them; there were many herds of 50 to 100, & I saw one, with, I should think, 500." This was a boon for the crew, since Darwin could shoot the animals and provide them with fresh meat, instead of the preserved "salt meat" they had brought with them. On the third day of the expedition, someone found a guanaco dead in the water, and the crew decided that even relatively recently expired guanaco beat their usual rations. The waterlogged carcass was "soon cut up & in the evening eat."

The river sliced through the narrowest part of a wide valley in the plains, and the road ran along the edge of the southern bluffs, the Santa Cruz visible as a thin ribbon of silver a half-mile off on the passenger side of the car. After about sixty-five miles of remarkably similar terrain, marked every ten or twenty miles by a sign for a new sheep ranch, a glorified horse trail dropped abruptly off the highway and down the cliff to a ranch called the Estancia Rincón Grande—

roughly, the "Big Curve Ranch." I decided to follow the side road and drove down to the farmhouse with tales of legendary Patagonian hospitality dancing in my head.

The estancias of Patagonia, for those who haven't read the classic Patagonian travel literature, are welcoming, enchanting places where hospitable ranchers sit around waiting for surprise guests in need of lunch, tea, and possibly a good night's sleep, to stumble in. Take, for example, this sheep rancher's quote from Bruce Chatwin's *In Patagonia*, the travel standard by which all Patagonian travel is measured: "'Look here, come in and let me cook you some dinner. Fancy finding this place on your own.'"

I pulled up behind a corrugated metal farmhouse and stopped at an outside garage where a man stood fiddling under the hood of an idling blue pickup truck. A row of neatly planted poplar trees, gold and orange in the fall, lined the house, which was done up in traditional farmhouse colors, white with green trim. I parked and the man stopped fiddling.

As quickly as I could, I explained what I was doing and asked if I could wander across the ranch's private property to go see the river. "Oh," he said. "You want to talk to the boss. Come on."

The ranch hand introduced himself as Carlos, and I followed him as he entered the house through a heavy swinging door. In a room to the left, I saw a flash of silver as a man sliced through a piece of meat. It was about lunchtime, and I half-expected an invitation.

Carlos did not stop, and led me down a narrow, poorly lit corridor to the rear of the house. He instructed me to wait and opened a door at the hallway's end. "There's a boy here to see the river," I heard him say. For some reason, the diminutive-sounding phrase he had used to describe me, *muchacho*, stuck in my head. I didn't hear the *jefe*'s response, but Carlos waved me in.

The office was small, dirty, and dark. The boss, an elderly man with tanned, weather-beaten skin and a silver beard, sat behind a desk in one corner. A young man I presumed to be his assistant sat behind a desk in the other corner. The boss was doing something that neither Darwin nor Chatwin mentioned in their romantic raptures about the freewheeling ranch lifestyle: paperwork. He looked up. He shuffled the papers in his hand.

"Yes?" he said.

His tone was neither friendly nor welcoming. It did not indicate any promise of lunch with the family.

"I'd like to see the river," I said. These words came out a bit more demanding than I intended—in translation, I'd lost much of my ability to qualify my statements. The boss looked surprised.

"What?"

I tried again. "The Beagle, with Charles Darwin and Robert FitzRoy, came this way. I'd like to see where they went."

"Oh," he said. He displayed zero curiosity about this exciting historical event that had taken place on his property. "Fine. Go ahead."

I lingered for a few seconds to see if a lunch invitation was forthcoming, but the boss sat back down and resumed his paperwork.

"Thanks," I said.

Carlos led me back out of the house. I got in the car and drove down toward the water, and parked at a wide spot in the road where it seemed like I'd be able to hike a short distance to the river bank. A few minutes later, while I stood contemplating the dry, chalky plains around me, Carlos drove up, waved, honked, and sped past on his way to tend the sheep. Even after the truck had disappeared over a hill, I watched the plume of dust rising in the air that marked its journey across the ranch. I set off walking, lost sight of the river for a few minutes, then stumbled over a small rise and down a gravel bank. Up close, the river looked about the same as it did in Piedrabuena. Still green, still moving fast. At ten miles a day, Darwin would have seen essentially the same scenery every day for two weeks. On the fourth day of the trip, he wrote, "The country remains the same, & terribly uninteresting."

Two weeks later, he wrote again, "The country & its productions remained equally uninteresting."

A few hours later, back on the main route, the setting sun dropped below the Corsa's small visor. Light streamed through the windshield and made it hard to pick out the worn path through the road. Rocks bounced and rattled off the bottom of the car, sounding metallic pings, like a squirrel firing pine cones at an aluminum roof. I slowed down to cross a metal-pipe bridge and saw an unmarked, car-width dirt path heading off to the right, in the direction of the water, which had disappeared behind a small ridge.

I had a few hours remaining before it got dark. I pulled off to see where the less-traveled road would take me, and it dropped directly down to the river (green, fast, etcetera) and emerged at a huge, flooded 90-degree bend in the water, part of a series of switchbacks that extended to the horizon, where the river swept behind a mountain and disappeared.

Near the end of his Santa Cruz expedition, Darwin wrote: "The river here was very tortuous . . . which sadly interfered with our progress." After following a relatively straight river for the entire day, this was clearly the first part where it had been anything approaching tortuous. Looking at a map later, it was actually the only part that was winding.

The thought slowly sunk in: Darwin walked here. Leaving the car next to a small patch of scrubby bushes, I decided to walk as well. A freezing wind gusted out of the northwest, and except for the ululations of the guanacos, it was very, very quiet. Bleached bones lay strewn across the ground: leg bones, jawbones, the ribcage of an ostrich, piles of femurs and tibias, even a bloodstained pair of guanaco molars.

The plain was absolutely deserted and likely unchanged in the last 180 years—if not the last 1,800 years. With little effort I pictured a group of bored, frustrated Englishmen hiking along the river's edge, hauling boats. Darwin walked out in front, sometimes up along the cliff, scouting and looking for guanacos to shoot. His boots would have scraped the tough yellow grass and sunk into the chalky soil, leaving strong prints in the ground. Rheas and guanacos would have watched warily from the hilltops, wheeled, and scattered as he approached. Dust would have swirled and settled on his jacket, his gun, and his hair. The rest of the crew would have trudged behind him, wishing they were back in the King's Head or Goat & Feathers or whichever old English pub they favored, cracking dry jokes about sporting pursuits and women.

I followed my imaginary Englishmen out along a 40-foot-tall bluff, toward a sharp hairpin turn in the river. The point revealed little except another bend, the kind of view that convinced FitzRoy that nothing lay ahead but more hard work. On a satellite map, or even from the top of the nearby mountains, he would have seen a thin strip of blue winding through the hills until dead-ending in a massive lake just a few miles from where he stood. But deprived of those navigational tools, and frustrated with the slow pace, he sounded the retreat. On May 4, 1834, Darwin and the captain walked westward as far as they could, turned their backs to the mountains as the sun set, and floated down the river the next morning. They never knew how close they had come to the river's source, and two days later, everyone was back aboard the Beagle.

Now I stood at about the same point, watched the s u nset, and thought about how rare it was to find a place preserved mostly the way Darwin saw it—a place that, perhaps because of its winds and lack of rain, had been bypassed by South America's evolution. While Rio de Janeiro had been built up and Port Desire had rusted away, the Rio Santa Cruz pressed on, unvisited, desolate, its banks lined with bones. Humans have never really figured out how to permanently settle in this land, making it one of the few areas on earth that still changes on a geologic scale. "The plains of Patagonia are boundless, for they are scarcely passable, and hence unknown," Darwin wrote. "They bear the stamp of having lasted, as they are now, for ages, and there appears no limit to their duration through future time." I had reached the ultimate destination for a modern traveler: the stereotypical unspoiled, rugged, nineteenth-century version of the South American outdoors. The kind of place that, if there's any romance at all in your adventuring soul, makes you sit up in your cubicle and take notice.

One, that is, that Darwin hadn't remotely enjoyed.

The more I read and traveled, the more it became apparent that Darwin hadn't necessarily derived pleasure from exploring unexplored wilderness—though the idea, like the *idea* of Patagonia, lent extra excitement to the places he did enjoy (like the relatively well-mapped Brazilian rainforest). The myth of Patagonia was all well and good while Darwin wasn't actually there, but when he was, when he really dug into that myth and lit out for the great open spaces, he found it dull. Instead Darwin reveled in novelty—the distance from home and the gulf of difference between the stately Shrewsbury countryside and the exotic South Atlantic. The real thrills—and the real education—came in interacting with people and places he couldn't find in England, which is why the businesslike trip up the River Santa Cruz, with its repetitive scenery, wasn't as much fun. And for me, as for any modern traveler, capturing that spirit of novelty and discovery requires an understanding not only of the destination, but also of home.

PART II
REVOLUTION

THE CHANGING PEOPLE ON THE EAST COAST

5: ENGLAND

One Last Frozen Image

I am determined & feel sure, that the scenery of England is ten times more beautiful than any we have seen.

—DARWIN, IN A LETTER HOME, JULY 18, 1836

There we British sat, poor grey sodden creatures, huddling under our grey northern sky that seeped like a rancid dishcloth.

—DOUGLAS ADAMS, *THE SALMON OF DOUBT*

ON A TYPICAL WELSH SUMMER DAY I hunched against rain and wind, burrowed down into the sodden collar of my jacket and thought back to an admonition I'd heard earlier that morning, "there's nothing like Wales when it's wet." Mist swirled, fog condensed, and beads of water dripped from my hair onto my nose. Two months after leaving South America, I was here to see where Darwin had come from and where he would return, never to leave again. A sharper contrast between the arid Patagonian plains and this leaking-sky climate would be hard to draw.

"Ah, drizzle," explained my English friend Nathan, pulling up beside me on the rocky summit of Snowdon, Wales' highest mountain. "It's more English than scones, drizzle."

One hundred seventy-five years before us, Charles Darwin stood somewhere near here and took in his last mountaintop view of Great Britain for five years. Even on a moderately blustery August day, he would have seen a beautiful panorama of jagged rocks sawing through saturated turf, deep blue mountain lakes pooled under soaring crags, and pastoral stone fences crisscrossing verdant pastureland.

"I can picture Darwin at the top," Nathan continued. "He would have stood here and said, 'I can't see shit.'"

Nathan Cooper was an old friend and, more importantly for my Darwin-studying purposes, an old friend from suburban Worcester with a taste for adventure. My trip up the River Santa Cruz had given me the idea that to really dig into South America I'd need a better understanding of England. Somehow this notion did not sound so crazy when repeated to someone like Nathan. I liked bouncing ideas off of Nathan because he'd say the first thing that came to mind and, usually, it was hilarious.

"Darwin," Nathan said, the first time I asked. "He's the guy with the beard, right?"

"Um," I said. "What else?"

"*Golden Hind*," he said, referencing an English exploring ship, albeit one that belonged to Sir Francis Drake and thus predated Darwin by about three hundred years. "And a man who is pathologically obsessed with worms." (This is true, by the way.)

He paused to collect more thoughts. "Also," he said, "they had a poll a while ago about who was the greatest Briton, and Charles Darwin was number two. Behind Churchill."

Nathan's wild speculation was not limited to Darwin. He had literally emerged from the mists and into my life several years earlier, on a frigid morning in Cusco, Peru, as I sat in a van with several other language school students on our way to backpack in the high Andes. While the rest of us had stuffed our packs for an icy adventure and were now huddled for warmth, Nathan strolled down the street—twenty minutes late, as the van was just about to leave—in a light fleece jacket, carrying an uncharged digital camera and an iPod in a backpack just large enough to fit a water bottle. Fortunately, he didn't have a water bottle, so there was plenty of room. Two days later, while wearing my extra jacket and watching me cut up food to share, he looked up at me and remarked, "Mate, you do come well-equipped on these trips."

I struggled for a moment for an adequate response ("umm"), failed to find one, and then, as often happened with Nathan, just laughed. He laughed right back at me, and I promptly decided that here was most definitely a friend worth keeping around.

For a few years, Nathan's job consisted of playing the FTSE—the English stock market—via Internet trading, so he wasn't required to be in any one location, which he took to mean he might as well be anywhere. Since our meeting and subsequent two weeks of traveling together in Peru, he had kept in touch through a series of hilarious emails that described chess matches with drug dealers in small-town Andean bars, sheep impressions from street touts in Bolivia (a particular hobby of his, trying to get people to imitate sheep), and a triumphant ride on the shoulders of the citizenry through the streets of Holland. My personal favorite was his citizenship in a self-declared new country in Western Australia. Apparently, he'd stumbled across some sort of Aussie tax-revolt and decided on the spur of the moment that he'd always wanted a second passport. ("I'd like you to meet his Excellency sometime," he told me. "If you become a citizen too, maybe next year we can have a soccer team.") A disaster of a planner who was relentlessly cheerful, Nathan lived by the motto, "Cheeky Not To," which he'd recite happily each time some bizarre challenge stared him down.

Nathan had now settled down in western England and worked, frighteningly, as a driving instructor. (Even more frightening, he appeared to be really good at it.) And it was only here, about an hour from the spot where Darwin grew up, that I began to understand why it really was cheeky not to.

Darwin was fascinated by the idea that he—a young man never considered extraordinarily smart or talented—had arrived at a very radical, very different new idea. Yet one of the fun things about evolution is how extremely logical it is. (As Thomas Huxley, one of Darwin's great defenders, recalled thinking, "How extremely stupid not to have thought of that!") Despite nearly two centuries worth of concerted attempts to kill it, evolution by natural selection survives because it is a triumph of common sense. And one of the questions I liked to ponder while I traveled was whether the theory had clarified itself in Darwin's head after a few eureka moments, or if the specific examples themselves were less important than the travel experience. Did Darwin learn the logic in England, then bring it to bear on examples as he found them in South America? Was this a product of education and place, where any similarly educated and disposed Englishman, introduced to the finches of the Galapagos, the rheas of Patagonia, and the fossilized remains of long-extinct giant sloths would have arrived at the same conclusion? (Say, for example, Darwin had stayed home, but someone had mailed him the fossils, finches, and rheas to study.) Or was there something about the travel itself, the length of time away from home, and the broad diversity of events—not just scientific, but cultural and personal—that turned him into an expert logician and allowed him to put all those specific examples in context?

I'm not sure Darwin himself could have answered this question. And he did occasionally consider it. In a delightful concluding chapter in his autobiography titled "An estimation of my mental powers" (a segment dedicated for the most part to lamenting that he no longer enjoyed poetry) Darwin evaluated his mental qualities like this: "I have no great quickness of apprehension or wit . . . My power to follow a long and purely abstract train of thought is very limited; I should, moreover, never have succeeded with metaphysics or mathematics. My memory is extensive, yet hazy. . . . I have a fair share of invention and of common sense or judgment, such as every fairly successful lawyer or doctor must have, but not I believe, in any higher degree." On the plus side, he wrote, "I think that I am superior to the common run of men in noticing things which easily escape attention, and in observing them carefully," and "From my early youth I have had the strongest desire to understand or explain whatever I observed,—that is, to group all facts under some general laws."

That early youth was spent unremarkably, perhaps, but in a way that prepared him well for life on the Beagle: outdoors, and in collecting plants and animals. Darwin's family hailed from Shrewsbury, a small country town three hours west of London by train. So while Nathan finished lecturing sixteen-year-olds on the merits of the roundabout, I went to visit some friends in Shrewsbury who lived a few blocks from Darwin's old haunts.

Although Shrewsbury has grown a bit since, Darwin was a country boy. He entertained himself in the great outdoors as a child, wandering alone along the Severn River, which wound just behind his home, and where he could dabble in hunting and fishing. He liked collecting all sorts of plants and animals from the outdoors, but became particularly smitten with collecting beetles, a mania apparently quite popular in England at the time. "I will give proof of my zeal," Darwin wrote in his autobiography. "One day, on tearing off some old bark, I saw two rare beetles and seized one in each hand; then I saw a third and new kind, which I could not bear to lose, so that I popped the one which I held in my right hand into my mouth. Alas it ejected some intensely acrid fluid, which burnt my tongue so that I was forced to spit the beetle out, which was lost, as well as the third one." A cartoon, drawn at the time by a friend of Darwin's, shows the young man wearing a top hat and riding a beetle, waving his collector's net above the caption, "Go it Charlie!"

Within half an hour of arriving in Shrewsbury, I found myself hiking through a pasture with a fishing pole and a hunk of Spam in hand. The Spam was for bait, a gift from my host, Rich, who was a rather expert angler. He'd lived all his life in Shrewsbury and knew all the fishing holes on the Severn River. (And probably, for that matter, most of the fish.) Rich had driven me out to the middle of a large pasture, parked his car in the shadow of a hayloft, and set off across the country, following a cow path toward the Severn. The way he picked his own route reminded me of hiking somewhere like the Monte Wood in Port San Julian, except the environment around us was radically different. In England, everything about the scenery, down to the almost stifling peacefulness, demonstrates man's ability to manipulate the environment. In South America, there was absolutely no question who wore the pants in the man-vs.-environment relationship. (Not that destructive humans weren't doing their best to catch up.) The human settlements I passed through felt fragile and slapdash. Even places like Rio de Janeiro, concrete-straitjacketed as they were, seemed overwhelmed by the absolute vastness of Brazil when I saw them from the airplane window. In Darwin's time, the contrast between North and South, stately countryside and untamed wilderness, must have seemed even greater.

We left off fishing late in the afternoon and drove back to town for fish and chips, beer in the pub, and a fervent discussion of the Liverpool soccer team. (Sample dialogue, between a soon-to-be-bride and her close male friend: "You told me in December that you'd be at my wedding no matter who was playing!" "Well, I didn't know they'd be at *home*.")

The next morning Rich had to work, so I wandered around the Shrewsbury town, touring the thirteenth-c e ntury castle, admiring the signs outside the pubs that granted the owners a license "to sell by retail all intoxicating liquor for consumption on or off the premises," contemplating the modern "Darwin Shopping Centre," and strolling through blackberry bushes along the Severn, which wound in slow, stately curves from our countryside fishing spot back through town and right up to Darwin's childhood home.

The Mount, as it was known, stood apart from the rest of the houses in town and was snugly walled off in a suburban neighborhood. (Darwin's father was a wealthy, popular country doctor, and he could afford to splash out on his estate.) A small plaque on the façade commemorated

Charles Darwin's birth and noted his "detailed observations of the Galapagos." The house itself had long since passed from the Darwin family, and now served as the local land valuation office, full of white-collar land valuers buzzing about attaching numbers to parcel maps. This was a curious thing to me: the country house that Darwin had purchased in his thirtiess, where he had lived and worked until his death, was now an English heritage site, perfectly preserved down to the furniture, paintings on the walls, and plants in the garden. I'd been to Down House, as it's called, a few days earlier, before coming out to Shrewsbury, and found there a steady stream of tourists, myself among them, wandering through, gawking at Darwin's rocking chair, and taking pictures of Venus flytraps in his greenhouse. While each house was significant to Darwin's life and personality, the relative importance attached to them—the touristy, well-preserved landmark house on the one hand, and the land valuation office on the other—seemed yet another reminder that Darwin the evolutionist superseded most other aspects of him.

I rang the bell at the front door of the Mount, and after a few moments, a pleasant, middle-aged red-haired woman let me in. "Sorry to keep you waiting," she said.

I asked if there was anything Darwin-related to see.

"There's really nothing left," she said, but then she added, "You can see the room where he was born."

I said I'd be delighted, and we walked upstairs. She led me toward a door, and as she was about to open it, a man in suit and tie walked out. "Is the Darwin room free?" she asked him.

He nodded and turned to me.

"But the doctor's not in," he said.

The woman opened the door for me. The room was plain white, with an uncluttered desk, a flat-screen computer monitor, and a small, framed display hanging on the wall that commemorated the site of Darwin's birth. It was empty and quiet. A window overlooked the front lawn and a mulberry bush that appeared old enough to have entertained a ten-year-old Charles. The woman watched as I did a quick turn around the room, and then walked me back downstairs and outside.

Nathan didn't have to work the next day, and he picked me up in the morning and drove me fifty miles to the west. In a packed parking lot at the base of Snowdon mountain, day-hikers in colorful parkas lined up to ascend the peak while less adventurous travelers skipped the exercise and boarded a smoke-belching rail car, which creaked its way up along an ancient track to the summit. Nathan and I walked, slipping on bathtub-smooth rocks and loose shale. Sheep clambered across landslides, bleating, and their wan cries drifted through the clouds. When we startled sheep they bounded away down the trail, long, cat-like tails swinging. This long tail surprised me; I pointed it out to Nathan. "That sheep has a really long tail!" I said.

"Oh, yes," he said, completely deadpan. "That's where the Welshman grabs hold." (Seriously, the English and their sheep jokes.)

We continued up through the fog. When we finally reached the summit, we were rewarded with a view of white sky and mist-enveloped, dripping black rocks. I thought of the young Darwin, his entire travel experience forged between here, Shrewsbury, and Cambridge, at the time a fairly suburban university town. He had gone backpacking near Snowdon after graduating from college in August 1831 because he hoped a few days of geologizing might help him to sort out his future. Darwin's family had secured him a job as a clergyman, and such a post would allow him to continue living in the country, hiking and casually researching natural history. It was a secure, safe future—and one that his restlessness, and desire for new opportunities for observation, wouldn't abide.

Darwin left the mountains and the fog behind in August 1831, at the age of 22, and returned to Shrewsbury, where he found a letter from his favorite college professor, J.S. Henslow, inviting him on a surveying voyage to South America. It took him less than a week to commit to traveling around the world for the next five years. I could easily see Nathan, and a great many Englishmen who found their home island simply too small to contain their desire to fill in the lines on the atlas, grasping at the promise of such adventures. It would be, obviously, cheeky not to.

But then we're still back at that question, about the power of specific examples versus the power of traveling. Darwin's insatiable curiosity and mania for observation were formed by his childhood and focused by his travels. He obviously needed the scientific findings from South

America to help think up evolution. But he spent five years on the Beagle and had plenty of learning experiences that had nothing to do with natural history. And some of them were just as influential in Darwin's mental development as his observations of the varying-beaked finches.

6: SALVADOR DA BAHIA

Beginning and End

> *If to what Nature has granted the Brazils, man added his just & proper efforts, of what a country might the inhabitants boast. But where the greater parts are in a state of slavery, & where this system is maintained by an entire stop to education, the mainspring of human actions, what can be expected; but that the whole would be polluted by its part.*
>
> —BEAGLE DIARY, MARCH 17, 1832

THERE ARE THINGS AN ENGLISH or American background can prepare you for when you travel in South America, whether you're traveling now or in the 1830s. Politics, for example. You can read about political history in books and then show up and get a pretty firm grasp on Hugo Chavez (in the modern era) or Juan Manuel de Rosas (in Darwin's time; more on the great Argentine general later). It's different, yes, but it's fairly easily understandable from abroad.

There are also things that it's hard to convey in books, but that are an unanticipated bonus pleasure to experience in person, like tropical rainforests. Most of us have a decent understanding of the concept of green, but you can show up in your first Brazilian forest and be blown away, as Darwin was, by how that particular green just *glows*, in a way that's almost challenging to the pastoral green of the English countryside or the subtle green of the drizzling Welsh mountains.

And then there are things that, no matter what you've read or talked about or heard, you can't prepare yourself for. For Darwin, this meant one thing in particular, and it's something that there's not much to compare to in the modern era. Darwin, despite everything in his abolitionist English education, and despite a long family history of activism, was not prepared to deal with witnessing slavery. Doing so, at least according to some biographers, may have been one of the single most important events in his entire life, and not just in terms of his personality, but also in terms of his work.

The Beagle's first stop in mainland South America was in Salvador da Bahia, on the northeastern coast of Brazil. For the first few days, Darwin delighted in the tropical scenery, hiking and collecting and scribbling gleefully in his journal about the glories of the rainforest. Then he hurt his knee and retired to his hammock. While Darwin rested, Captain FitzRoy went out to see the town, and when he came back, he told Darwin about having visited a great slave-owner. FitzRoy, according to Darwin, "defended and praised" slavery and told Darwin that he had asked many of the slaves he had seen if they wanted to be free, and all said no. They were happy as slaves, FitzRoy said. Darwin shot back that those answers weren't worth much when asked in front of the master of the house, and FitzRoy threw a tantrum, saying that since Darwin did not believe his word they could no longer live together and that the naturalist would have to leave the voyage. The other officers, more accustomed to FitzRoy's notorious hot temper, grabbed Darwin and invited him to eat with them, and soon enough, FitzRoy apologized and invited Darwin back.

That week, Darwin had his point made for him by proxy: There were Englishmen abroad in almost every large port, either living in town or on board other English ships in the harbor, and they generally came onboard to dine and converse with the gentlemen on board the Beagle, Darwin included. One of these men, the captain of a British warship called the Samarang, visited soon after the slavery argument. "Cap Paget has paid us numberless visits & is always very amusing," Darwin wrote. "He has mentioned in the presence of those who would if they could have contradicted him, facts about slavery so revolting, that if I had read them in England, I should have placed them to the credulous zeal of well-meaning people: The extent to which the trade is carried on; the ferocity with which it is defended; the respectable (!) people who are concerned in it are far from being exaggerated at home."

Paget told Darwin and FitzRoy of the atrocities and tortures he'd seen and quoted a slave he'd talked to as saying, as Darwin related it, "If I could but see my father & my two sisters once again, I should be happy. I never can forget them."

"Such was the expression of one of these people, who are ranked by the polished savages in England as hardly their brethren, even in Gods eyes," Darwin fumed that night in his diary. Turning his rage to FitzRoy, but hesitating to criticize him by name, he recorded an anonymous jab at the captain. "From instances I have seen of people so blindly & obstinately prejudiced, who in other points I would credit, on this one I shall never again scruple utterly to disbelieve: As far as my testimony goes, every individual who has the glory of having exerted himself on the subject of slavery, may rely on it his labours are exerted against miseries perhaps even greater than he imagines."

Jean, my forest guide in the Tijuca Forest near Rio de Janeiro, had mentioned Captain FitzRoy's pro-slavery sympathies while we hiked in the forest. "The Captain," he asked me. "What was his name?"

"FitzRoy?"

"Yes, FitzRoy," Jean said. "He talked a lot of bullshit."

In the years since Darwin's visit, Bahia has morphed from the center of Brazilian slavery into the best spot in

Brazil to celebrate and study African culture and the legacy of the slavery that Darwin so abhorred. The English had essentially forced the end of the slave trade in Brazil by blockading harbors and attacking slaving ships; I wondered if Darwin's impassioned anti-slavery remarks were still remembered. The descendants of slaves make up an estimated eighty percent of the population in Bahia, and their culture—food, religion, and most especially music—is the city's biggest tourist draw. Stepping out of the airport terminal at Bahia, I was greeted by a group of black women in flowing white robes operating a small food stand outside where they sold palm hearts fried in dendê oil.

Modern Salvador da Bahia is a sprawling city of more than two million inhabitants, and the jungle that Darwin enjoyed—"The town is fairly embosomed in a luxuriant wood," he wrote—has been cut away to well beyond city limits. Government-sponsored billboards lining the main roads cheered on growth and development. Hard-hatted workers posed in front of oil wells and manufacturing plants; "Bahia is growing," the caption promised. "Bahians, too."

I checked into a hostel in Porto Barra, a block from the beach at the crux where the Atlantic Ocean met All Saints Bay. The hostel décor screamed tropics: a canary yellow exterior lined with cherry-red hibiscus, bright orange paint in the kitchen, and sunburned European tourists napping shirtless in hammocks strung across the common spaces.

I had arranged to meet a Salvadoran friend-of-a-friend named Silas Giron there that evening, and he arrived around 8 P.M. after getting off work at a local music store. He looked nothing like I had imagined—I had pictured someone short, dark, and athletic, like all the people I'd seen in Rio de Janeiro and on the streets of Bahia as I drove from the airport to the hostel. Instead, Silas turned out to be an illustration of Brazilian diversity: tall and waif-thin, with mocha skin and long, dreadlocked blond hair that he wore tied up in a net. He played the guitar and talked seriously about his musical influences, which represented a typical modern Bahian mixture of samba, swing, reggae, rock, and pop, and seemed to owe particular inspiration to legendary Bahian-born guitarist Gilberto Gil, who had recently been made Brazil's minister of culture. Although his grandfather was in the hospital undergoing chemotherapy—which he didn't tell me until later—Silas offered to help me as best he could. Darwin, with the very last line of *The Voyage of the Beagle*, made a brief argument in favor of travel for young naturalists, concluding that "Traveling ought also to teach him distrust; but at the same time he will discover, how many truly kind-hearted people there are, with whom he never before had, or ever again will have any further communication, who yet are ready to offer him the most disinterested assistance." Silas, for me, was one of those kind-hearted people.

After we had introduced ourselves and chatted for a while, Silas suggested we go for a walk through Porto Barra, the hostel's quiet beach neighborhood a few miles south of the city center. The waterfront extended to a sharp point marked with a lighthouse, and sandy public beaches curled away from the point. We walked along the promenade overlooking the bay-side beach, which was fairly crowded even at close to 10 P.M. The groups of people lounging, sitting on towels, even trying the water, were diverse—mostly black, but also whites and those of obviously mixed race. I asked Silas about the diversity on the beach, as it often seemed to me that Brazil had succeeded admirably in integrating different races. He thought for a moment. "That's one of the reasons I like this beach," Silas said. "Everyone comes here, all different kinds of people."

Silas hesitated to talk about racism in English, worried that it would quickly strain his vocabulary. Brazil is not without many of the same lingering racisms found in the United States, with blacks having higher incarceration rates, lower incomes, and fewer college degrees. Turn on the television in Brazil, and you're likely to see popular soap operas featuring mainly light-skinned actors and actresses. But although he acknowledged the complexity of such problems, Silas did seem to think that racism in Brazil was rarely overt. "Here in Salvador, everybody is very proud of our black influences," he told me. "It seems to me that most Brazilian people try hard not to demonstrate any sort of racial prejudice."

I asked Silas how Darwin's anti-slavery rants played here, but he hadn't heard of them. So as we concluded our walk, Silas suggested I come by his university the next morning to meet his history professor.

When I woke up the next morning, I wrote down some questions for the professor, Carlos Eugênio Líbano, and ran them through an online translator that promptly rendered them meaningless in Portuguese. ("Knows surplus the visit of Charles Darwin the Bahia in 1832?")

I met Silas in a patio in the University of Bahia's Philosophy and Life Sciences campus a few minutes before his class was scheduled to get out. "Don't you need to be in class?" I asked as he came out to greet me.

"No," he said, thinking it over. He sighed. "I should go more often, but it is very hard." He explained that he often worked late, or stayed up late playing music, which made attending his 9 A.M. classes difficult.

It was a beautiful campus, on a hill overlooking the ocean, with several shady viewpoints surrounded by the kind of greenery that sent Darwin into ecstasies. Silas helpfully identified one small, shaded area, on a tiny promontory overlooking the blue ocean, as his favorite spot in Salvador to smoke a joint. "You should bring your friends back there," he said. "I think they will really like the view." He seemed to know most of the students crowding the patio to smoke, talk, and eat, which made our progress toward his classroom slow.

Eventually, Silas led me inside a small, typical university classroom, where his professor was wrapping up his lecture. Then he started to take roll, and as I stood at the edge of the door, Silas quietly slipped back into the classroom with a cheery "Here, professor!" when his name was called. After class a small crowd clustered around the professor's podium, and as I edged toward the front of the room, I felt like I was getting an audience with the Pope, or maybe Santa Claus, until someone in front of me vacated the line and I was next. The professor looked up at me expectantly, waiting for me to give my name so he could put an X next to it and mark me present. Silas jumped in.

"Professor," he said, "this is the friend I told you about, from the United States. He is researching Charles Darwin and his time in Bahia."

The professor raised a spiky black eyebrow in my direction. "Your questions," he said in Portuguese. "Speak."

I hastily unbuckled my backpack and handed my sheet of questions to Silas, who read the first one. "What do you know about Charles Darwin's visit to Bahia?"

The professor said he didn't know much about it. "I am not an authority on Darwin," he said. (His area of expertise was in studying slave life, specifically the acrobatic martial arts dance called capoeira, in the time of slavery.) "You can look up the microfilm, though," he added. "What dates was he here?"

"The end of February, 1832."

He shrugged, and indicated we should walk with him, Silas simultaneously keeping an eye on me while doggedly pursuing the professor through the crowd of students. I trailed behind both, doing my best to understand the rapid-fire Portuguese being spoken. Sometimes Portuguese sounds so much like Spanish that I thought I could understand, particularly after I had learned a few of the key differences. Sometimes, it really doesn't—this was one of those times. "How much did Europeans influence the abolition movement in Brazil?" Silas asked on my behalf.

"Very much," the professor said. "The English basically ended the slave trade by attacking Portuguese slaving ships that were leaving the harbor at Bahia."

"Darwin was an abolitionist," I told Silas as we both jogged along, now through the bright sunshine outside the classroom, "and he wrote about how bad slavery was in Bahia. His book was very widely read in England. Can you ask if this had any effect on attitudes there and here?"

He repeated my question to the professor, who by now had donned a heavy pair of aviator shades. I found him intimidating, although Silas seemed to find him as approachable as a smiling Labrador retriever.

"Darwin had no effect while he was here," the professor said. "But when he got home, his writings had a very big effect. He said he would never visit a country that had slavery again."

The professor turned away to purchase a small cheese-infused piece of bread from a food trolley and started to wolf it down while chatting with a colleague. Before Silas could grab him again, he rushed back to class.

"I really like him," Silas said as we watched him stride away. "He is very funny."

Europeans who refused to visit or move to slave-owning countries did, at least, play a role in slavery's demise. Although British influence had led to an end of the trading of slaves in 1851, slavery itself continued. Without new slaves arriving from Africa, however, Brazilian farmers began to look elsewhere for workers—and, along with much of the rest of South America, looked to European immigrants to fill the plantation workforce gaps. When these potential immigrants refused to go to a country where slavery existed, plantation owners began to think differently. By the time the Brazilian emperor ended slavery in 1888, he enjoyed almost unanimous popular support.

Darwin wrote more about the evils of slavery, and, as if to emphasize his feelings, he placed his most impassioned argument just before the conclusion in the published version of *The Voyage of the Beagle*. He first set out the horrors he had witnessed or heard of—beatings, whippings, families forcibly separated, and the moans of tortured slaves. Darwin added that he would not have mentioned these had he not "met with several people, so blinded by the constitutional gaiety of the negro as to speak of slavery as a tolerable evil." Then he summarized arguments for slavery and provided his own counter-arguments in the same meticulous way in which he later documented the case for natural selection.

For a few crucial pages, and at a crucial time in his narrative, Darwin completely sets aside everything he's famous for—geology, botany, biology—and focuses on the political issue that animated much of his life. Darwin biographers Adrian Desmond and James Moore argue that abolition wasn't just a passionate side-cause for Darwin and that it actually influenced his whole career, most particularly Darwin's second major evolution-related work, *The Descent of Man*. They argue that those last few lines in *The Voyage of the Beagle* were in fact a calculated argument inspired by Darwin's anger at a travelogue published by his close friend Charles Lyell that appeared to accept slavery in the United States. "Darwin retorted by publishing details of the most revolting tortures he could remember," Moore and Desmond wrote. "He lit the fuse buried in his notebooks and exploded against the 'sin' of slavery. Never again would he express himself so thunderously to the world."

Darwin's thunderous conclusion, written long after he had returned home and only for an edition of *The Voyage of the Beagle* published in 1845, still called up all the righteous wrath he had felt in FitzRoy's cabin thirteen years before: "It makes one's blood boil, yet heart tremble, to think that we Englishmen and our American descendants, with their boastful cry of liberty, have been and are so guilty."

7: LA PLATA

In Darwin's Muddy Footsteps

> *The other day we landed our men here & took possession at the request of the inhabitants of the central fort. We Philosophers do not bargain for this sort of work and I hope there will be no more.*
>
> —LETTER TO A FRIEND, AUGUST 18, 1832

NO ONE TRAVELS TO BE BORED. But without the structured to-do lists of home, it's an inevitable part of any trip. I'd guess that boredom is the most common, most universal of all traveling emotions. To combat it, travelers review and relive their few moments of glorious excitement by writing in journals (Darwin) or blogging for hours in Internet cafés (not Darwin). They think deep thoughts, and ponder philosophical and political questions, as Darwin did about slavery while lying indisposed in his hammock in Bahia. Or if they're like my English friend Nathan, they lash against the constraints of inactivity like netted fish. Traveling with Nathan was never boring.

The tiny Atlantic coast country of Uruguay is boring, in the nicest sort of way. It's clean, peaceful, tranquil, and scenic but not remarkably so. For Darwin, who spent several months wandering around on the north banks of the Rio de la Plata, Uruguay was at best a decent opportunity to practice his Spanish. "I never saw so quiet, so deserted looking a place," Darwin wrote in his journal, about the eastern part of the country. "After pacing for some weeks the planck decks, one ought to be grateful for the pleasure of treading on the green elastic turf, although the surrounding view in both cases is equally uninteresting."

For Nathan, I think, three days in Uruguay nearly killed him.

I didn't make it to Uruguay while following Darwin. But Nathan and I had been there before. When I reviewed that trip later, as I read *The Voyage of the Beagle* for the first time, the parallels between our three days bouncing along the Plata and Darwin's experiences leaped out at me almost the same way the story of the Darwin's rhea had.

I had met up with Nathan in Buenos Aires, where I had just landed, and where for the last few weeks he'd been an active conspirator in the nightlife at one of the most scandalous and legendary youth hostels on the continent.

Although there was a particular Spanish language instructor he had in mind, the entire city had caught his eye.

Buenos Aires is an incredibly aesthetic city: beautiful people in beautiful clothes eating beautiful food from beautiful dinner plates in beautiful buildings. Walking the streets in the morning I would be passed by tall, thin businessmen and women strutting off to work in fashionable suits. The city seemed to be populated by energetic 30-something business people straight out of Hollywood casting central, the kind of people you always see working in advertising firms in romantic comedies. To my alarm, I discovered that Nathan had bought new shoes, a new jacket and a few new dressy shirts, He had also joined a gym and now kept up a daily exercise regimen that included brisk jogging around the park, and he was concentrating enough on his Spanish lessons (that instructor—not only pretty and fashionable, but a good teacher!) that he'd chucked his charming disregard for the language out with his old clothes. And then he'd turned into a zombie by trying to skip sleep for three weeks.

Darwin, who spent very little time in Buenos Aires, sounded something like everyone else in the hostel. (Or rather, they sounded like him.) "The whole town has more of an European look than any I have seen in S. America," he wrote after one trip. ("The Paris of South America!" modern guidebooks proclaim.) Like my hostel mates, Darwin found little to do except roam the streets admiring the finely dressed women. "Our chief amusement," he wrote to his sister Caroline, "was riding about & admiring the Spanish Ladies. After watching one of these angels gliding down the streets; involuntarily we groaned out, 'how foolish English women are, they can neither walk nor dress.' And then how ugly Miss sounds after Signorita; I am sorry for you all; it would do the whole tribe of you a great deal of good to come to Buenos Ayres." (Check.)

The problem with youth hostels is that they don't really want you to turn into a permanent resident. So when the Buenos Aires hostel staff chose to kick Nathan out for having overstayed their three-week limit, and we needed to go somewhere for a few days to relax, we chose nearby Uruguay. It turned out to be a little more relaxation than Nathan, at least, had sought, which is why three days later, within hours of our return to Buenos Aires, we found ourselves in a small, dimly lit underground chamber in the theater district, sitting with three groups of women who were having bachelorette parties and two groups of men who were having bachelor parties, all of us—and most especially Nathan—facing an eight-foot circular pool filled with mud and two wrestlers in thong bikinis that were also, rapidly filling with mud. Darwin probably would have done the same.

In 2005, a Uruguayan pop music star named Jorge Drexler was nominated for an Academy Award for writing a song called "Al Otro Lado del Rio" for the soundtrack of the movie *The Motorcycle Diaries*. When Drexler asked to sing his own song at the awards show, the producers told him he didn't have enough star power and that they had decided the song would be better handled by Hollywood's most recognizable Latino, Antonio Banderas. Banderas performed and Drexler ended up winning the award. His rebellious acceptance speech was to sing, very sweetly, a few lines from his song. The entire affair was huge news in Uruguay but went mostly unnoticed in the United States. A few days later, Jorge's brother Daniel Drexler told the *Wall Street Journal*: "We got excited when they mentioned the word 'Uruguay' on *The Simpsons*, even though they pronounced it 'you are gay' and made a joke out of it."

Uruguay may be—and I say this with the utmost affection—one of the most irrelevant countries in the world for a twenty-first-century North American. It's not in the Middle East or Europe or Asia, it doesn't participate in or approve of the war on terror, its environmental footprint is probably around a quarter that of Los Angeles', its socialist leader isn't bombastic, its two major sports are soccer and rugby, and it has no major tourist sites, ancient ruins, iconic waterfronts, famous cuisine, or (sorry, Jorge) identifiable rock stars.

"Whoever has seen Cambridgeshire, if in his mind he changes arable into pasture ground & roots out every tree, may say he has seen Monte Video," Darwin wrote after climbing a hill near the Uruguayan capital.

Or, as Nathan said, while slinging his backpack onto a bunk bed in an empty hostel dorm room in Montevideo, "I think Uruguay needs to work on its PR."

The hostel was a huge three-story house with room for forty-eight guests in which we appeared to be the first visitors. Nathan, having dropped his bag, scanned the room for signs of intelligent life or at least signs of something promising hedonism and, finding none, plunked down moodily on the bed.

"Well," I said, unloading my own pack in the opposite corner of the room, "what can you really say?"

"Uruguay!" Nathan cracked, perking up briefly. "It's not Paraguay!"

And yet Uruguay was far more like my home in California than I'd ever have imagined. Montevideo and San Francisco are almost equally distant from the equator (one north, one south), and the average temperatures were within a few degrees of each other. Both Uruguay and California had been colonized by the Spanish, who had nearly annihilated the local Indian presence (leaving the few remnants so scattered and blended by intermarriage that tracing any kind of distinct indigenous history is nearly impossible). They had won independence from Spain four years apart, Uruguay in 1828 and Mexico, which California was a part of, in 1832. Both had carved the land into ranches for wealthy veterans, who had then fought losing battles to keep immigrant squatters from further dividing it. Both had, at one time, relied almost exclusively on trade in cattle—a shorter-lived period in California history, which changed abruptly when gold was discovered in 1848 and scofflaw Americans began to move in. Both regions spawned a breed of fiercely independent cowboys, Uruguay's gauchos and California's vaqueros, famed for their horsemanship and rugged lifestyle. (One article in the *New York Times* suggested that George W. Bush liked visiting Uruguay because there was a nice ranch house for him to visit that reminded him of Texas. Plus the non-bombastic socialist leader thing.)

Finally, triumphantly, both Uruguay and California had somehow ended up with suburban tract-home blight on top of many of those pastoral ranches. "Going north from Punta you will pass some of the finest residential areas in South America," one Uruguayan tourist website proclaimed. "This area of the Maldonado Region is a 'must see' for tourists and visitors. Golf and tennis clubs can also be found in this vicinity."

Golf! Tennis! Uruguay!

Montevideo was a comfortably nice place, with tree-lined shady streets, old Victorian-looking houses, and an air of quiet retirement about it. Nathan and I strolled around for a while and ended up reclining on a grassy lawn at the edge of the Plata, basking in late afternoon sunshine and a sea breeze, lazily watching a few dozen men on a rocky breakwater as they pulled flashing silver fish out of the brown water. Shrewsbury-native Darwin may have found this a "great air of wealth & business," but for us, it looked a lot like a bigger, more populated suburb of the rest of the sleeping country. A rugby team was practicing on the lawn next to us, running formations, tossing the ball from one end of the line to the other as they ran from end to end, and I felt a soothing sense of home. The grass and the fishing and the sports could just as well have been set in any marina in the San Francisco Bay.

But while I slid down the grass and relaxed into the slow-paced, somehow familiar Uruguayan rhythm, Nathan ground his teeth and openly despaired of ever finding seediness and wild hedonism again. "This is a capital city," he said, through set jaw. "There must be nightlife somewhere."

In Montevideo, Darwin had attended a grand ball to celebrate the reestablishment of the president. "It was a much gayer scene than I should have thought this place could have produced," he wrote.

While I was visiting Nathan in England—the first time I'd seen him since our paths had diverged in South America and I'd discovered Darwin—we compared notes and talked about the naturalist's take on Uruguay. I read the bit about the grand ball out loud. "He found a party?" Nathan asked, incredulous. "Where? How? Can we find it still?"

The partygoers, Darwin reported, were finely dressed —the women in particular—and he expressed his amazement that even the lowest classes of society were allowed into the theater to observe the dancing. "And nobody ever seemed to imagine the possibility of disorderly conduct on their parts," he marveled. "How different are the habits of Englishmen, on such Jubilee nights!"

A determined Nathan, ready to uphold his country's reputation, burst out onto the Uruguayan streets that night. The hostel operator looked surprised. "We lock the doors at midnight," she said quietly.

For a while, we blundered around the dark residential streets. We found a few restaurants, some cheerily lit pizza joints, and some shuttered office buildings. As Nathan got increasingly worried that there really might not be any bars in all of Montevideo, we reached the older part of town, with narrow pedestrian-only streets that reflected the lights from a few open doors. Suddenly, we came to a swinging green sign for "The Shannon."

"Irish pub," I read.

Nathan swerved. We had stumbled into the capital's bar sector, with four or five different bars serving up music and drinks, and Nathan, back in his element, looking like he had finally found his ideal place to settle down, was hesitant to leave. "It's like Irish bars everywhere," he said, gratefully gulping his beer. "Guinness, dark wood, no actual Irish people. It's absolutely wonderful."

We sat wedged in the corner at a dark table, with Uruguayan waiters constantly tripping over my feet as they passed by on their way back to the bar. Nathan looked pleased. He later suggested that Darwin might have had his evolution epiphany while looking for a drink in Montevideo. "It's survival of the fittest," Nathan said. "It's like some male of some soon to be extinct species crawling through the jungle in one last-ditch, futile attempt to find a female. Maybe it could have been Darwin's eureka moment."

In October 1833, Darwin boarded the Beagle anticipating that it would soon be leaving Uruguay for good. But he was disappointed to hear that the navigational charts had not been finished yet and the ship would stay in Montevideo until December. Determined, then, to do something with his time, he decided to ride to the river marking the western boundary of the country. A few days later he arrived in one of Uruguay's original settlements, Colonia del Sacramento, and found much of the town in ruins. The ancient church at the center of town had been destroyed—it had been used as a powder magazine and was struck by lightning. Darwin walked along the half-demolished walls for a while, pronounced it a "pretty" little town, and left the next morning.

Nathan and I arrived in Colonia early in the afternoon on a bus from Montevideo and found that it was now one of the most pleasant, sleepy towns on the continent. Tree-lined, broken sidewalks funneled visitors down toward the waterfront where the Rio de la Plata lapped gently at a rocky breakwater. The town looked like it hadn't been touched much since Darwin strolled around admiring the ruins. Walking down the main street, we ran into a hard-partying Australian whom we had last seen days earlier at a Buenos Aires pool table, standing in a haze of cigarette smoke with cue stick in one hand and a liter bottle of Quilmes beer in the other at something like six in the morning. He had been in Colonia now for three days.

"What have you been doing?" I asked, looking to the end of the street and thinking it was an awfully small place to keep him entertained for a week.

"Detox," he said. "I go to bed early, I sleep late, I go for a walk. It's just such a nice place to relax after all the partying in Buenos Aires."

"When are you going back?"

"Don't really know," he said. "Maybe by the end of the week. I've gotta get back there at some point."

He agreed to have lunch with us in a shaded sidewalk café on the main drag. Nathan dove into the national dish, steak with ham, bacon, and an egg, and we leaned back into our chairs and felt laziness overcoming us. Colonia was a brilliant place to take a leisurely stroll, and everyone was doing it; families, groups of teens, elderly couples, almost all of them carrying carved, gourd-like containers and hot-water thermoses for their *maté*, a boiled-grass tasting tea-like concoction which they drank through metal straws. It was late in the afternoon, and the tea hour had arrived.

Maté is an addiction widely popular in Argentina and Uruguay, and its popularity is growing among natural-food enthusiasts in the United States. To prepare it, you pull out a small bag of the herb and pack it into the bottom of your "calabash," usually a hollowed-out gourd made for *maté*. Add hot water and a metal straw, pass it around, and you've got everything you need for a pleasant afternoon. South Americans drink it straight, without sugar, although it takes quite some time to adjust to the flavor of freshly cut lawn. Darwin, by the end of his journeys through the Pampas, seemed to have grown quite fond of his daily "mattee."

We saw people slowing up to rest against store walls, pulling thermoses out of backpacks, sharing their tea with friends. It was peaceable, friendly, and relaxed, and after lunch we meandered down to the harbor where we watched the sunset over the still water and admired the sailboats. Groups of people in shorts and sandals were out taking their tea on the decks of the boats, a father and his son were fishing from the end of the pier, and the Australian sat looking at the purple water, feeling the sun and the wind on his face with an expression of contentment, like he was telling us, "See what I mean?"

I did; Nathan did not. But now, he had a remedy.

For days, he had been scanning the papers looking for nightclub news. Back in Buenos Aires, this very night, he announced, was a mud-wrestling show. And, he said, after an almost incredible amount of effort (for him), he had secured two tickets. We'd just hop on the ferry, ride back to town, and finally get some of that culture we'd been missing in our relaxing country retreat, which is what we did.

Later, as I mulled over the mud-wrestling show, and Nathan's enthusiasm for it—expressed in his usual excited terms as we raced back toward the ferry—I tried to imagine what Darwin would have said. In Uruguay, in our boredom, in Nathan's search for bars, and even in watching his reaction to the nightlife situation, there was a nice parallel: we were following in Darwin's footsteps by not following in them, without knowing that's what we were even doing. By desperately seeking a bar, we were reacting to our own boredom; Darwin felt the same way, even if his choice of ameliorative activity (riding around on a horse, pursuing lizards to throw) was different. Now, as we raced through town to get back to the ferry that night, we were chasing the same idea of señoritas (the, ahem, pure Platonic form of señorita) except we were going to get the modern traveler's version, as part of the cultural component of this Darwin quest.

I later shared my conclusions with Nathan. "I suppose," I said, "that a mud-wrestling expedition could count as research. We went to see if the Spanish Ladies still held their charms when undressed and caked in mud."

Nathan grinned.

"Your Darwin," he said, "is a bloke I'd like to meet."

8: TIERRA DEL FUEGO

Darwin Visited (Near To) Here

> *I shall never forget how savage & wild one group was. Four or five men suddenly appeared on a cliff near to us. They were absolutely naked & with long streaming hair; spring from the ground & waving their arms around their heads, they sent forth the most hideous yells. Their appearance was so strange, that it was scarcely like that of earthly inhabitants.*
>
> —BEAGLE DIARY, JANUARY 20, 1834

OF ALL THE PLACES I VISITED, Ushuaia, the biggest city in Tierra del Fuego, most aggressively marketed itself as a Darwin destination, based on four key realizations:

1) Darwin visited (near to) here
2) Gringo tourists are interested in Darwin

3) Gringo tourists do not understand much about Darwin
4) We can make money by presenting Darwin to Westerners

Which is basically how you get "The Adventure of the Beagle," the musical.

Again: "The Adventure of the Beagle," the musical, or, as it's called there, *El espectáculo del fin del mundo*, "the show at the end of the earth," a production of the tourist-friendly *Centro Beagle*.

I arrived at the Beagle Center an hour before the show was supposed to start and bought a ticket from a young man in a sailor hat that said "Beagle" on it. I asked if he knew anything about Darwin, and he said not much, and could I recommend a book? I suggested *The Voyage of the Beagle*, and, pleased, he handed me a ticket and a paper labeled "boarding agreement" and waved me into the waiting room.

The lobby was huge and featured a number of wooden tables arranged around the aft end of a scale-model replic a of the Beagle. It was empty, so I grabbed a bar stool and read through my "boarding agreement" while the lobby speakers blasted a looped recording of Handel's "Coronation Anthem, Zadok the Priest." The boarding agreement had two clauses, "purpose of the journey," which was mostly correct in outlining the purpose of the real Beagle's mission, and "on board rules," which let me know that the "(1) captain is the absolute and final authority on board" and "(2) smoking, taking pictures, video, and/or audio recording use of cellular phones and other electronic devices is strictly forbidden." For four repeats of the Coronation Anthem (you've probably heard it before; it's also famous as the anthem of the European "champions league" club soccer tournament), I sat alone and wondered if they would perform the show for a solo, although highly enthusiastic, spectator. Would I be allowed to participate? Pause the actors mid-performance to ask questions? How gruesomely, fascinatingly awkward would *that* be?

But ten minutes before show time, a group of twenty or thirty cruise ship passengers arrived, with a tour director in tow. They seated themselves at a long table and ordered pre-performance Argentine wines. As waiters brought the wine, a few actors in what some costume designer evidently imagined to be a nineteenth century sailor outfit—with their wide-brimmed floppy baker's hats and silver-blue neckties, they looked more like Parisian pastry chefs—emerged and pretended to swab the decks.

Before the violins could strike up a seventh chorus of the Coronation Anthem, the cruise ship passengers rose in unison at a signal from their tour director and shuffled through a curtain in the back of the Beagle replica. I followed them onto the ship, where the rising part of the rear cabin had been converted into seats. I climbed up and found one near the top, looking down at the deck planks and the sail-less main mast. My view also encompassed Tierra del Fuegian scenery, in this case glaciers made of crinkly white fabric draped over a metal framework. The ship sat in the middle of a big-top-style tent, and the black walls, with inset starry lights, rose high overhead.

Then everything went dark. The audience went quiet.

The production began with a small video screen showing—what else?—Darwin as an old man in his study, recounting the origins of the voyage in a frail, Argentine-accented voice. And then, suddenly, our hero appeared in the flesh, emerging from a door in the Beagle's fore cabin dressed in a long corduroy jacket, long red pants, and a pink shirt, and carrying what looked to be a rusty red suitcase emblazoned with flowers. The actor's costume appeared to be loosely based on a portrait of Darwin at age thirty-one, wearing a brown shooting jacket, with a blue vest underneath—the cut of the actor's jacket was similar, even if the color was off. The actor himself had long, straight hair, dyed blond with visible dark roots, and worn in an unruly mop. Synthetic muttonchops were glued to his cheeks.

Soon, the squadron of sailors that had earlier been pretend-swabbing the deck appeared, and then an appropriately costumed FitzRoy and another officer, and then, as Monty Python might say, things got silly. They broke out in song, in English. The lyrics, transcribed on the "boarding agreement," read:

> We'll fight the roaring seas
> We shall face no defeat
> All across the Seven Seas
> The Beagle will succeed.

When the sailors finished singing and stomped off to work, Darwin and FitzRoy took center stage for a duet about searching for truth (in Darwin's case) and the work of the Lord (in FitzRoy's case). The lyrics—Darwin: "I'll listen to the calling of the Earth," FitzRoy: "uncover all of nature's divine perfection and more"—played on two falsehoods, the first being that Darwin had any kind of coherent conception of his theory of evolution by natural selection before, or even really during, the Beagle's voyage. (Mostly, he had inklings that something wasn't right with the traditional explanation for the origins of life, which held that all species were created exactly as they were and did not change.) The second untruth was that FitzRoy wanted Darwin along to prove the literal truth of the Bible.

The musical, in fact, continued to hammer that point home, portraying FitzRoy as an overbearing fundamentalist blowhard, unwilling to tolerate dissent on religious matters. It staged Darwin as a tormented evolutionist, torn between his friendship with the captain and the new scientific truths he was discovering. If this theme wasn't evident from, say, the twenty-foot-tall dancing sloth fossil that sang to Darwin that "you can try to deny what your eyes meet . . . but think you fool, don't be a mule . . . I am as real as these bones," the wailing solo that Darwin sang near the end removed any remaining ambiguity:

> There's no way to go on
> And there's no turning back
> Nowhere to run
> Nowhere to hide
> I'm torn inside.

In fact, Darwin did complain frequently about his insides in Tierra del Fuego—with the bad weather and choppy seas, he was constantly seasick.

I knew that the play's take wasn't particularly historically accurate and wasn't particularly surprised, since these are fairly easy assumptions about Darwin and the Beagle voyage. In fact, it's probably easier to present this material —see point 3, above, "gringo tourists do not understand much about Darwin"—than to risk audience cognitive dissonance by trying to tell what actually happened.

Probably more exciting, too. Because in real life, the drama-worthy conflict on the Beagle was man versus the elements, which is hard to render in musical format. (Although if you've got twenty-foot-tall sloth bones singing, why not the snow gods?) Far from being a biblical literalist, the real FitzRoy wanted Darwin to study science. "Anxious that no opportunity of collecting useful information, during the voyage, should be lost; I proposed to the hydrographer that some well-educated and scientific person should be sought," FitzRoy wrote in his *Narrative*. He had outfitted the ship with the very latest in scientific equipment and took a strong interest in Darwin's discoveries. Although FitzRoy later became a staunch critic of *The Origin of Species*, standing up at one meeting with Bible in hand to say that he "regretted the publication of Mr. Darwin's book," there is little evidence to suggest he was particularly religious while on board. Darwin remembered that FitzRoy "became very religious" *after* the voyage and that in his own case, "I was quite orthodox, and I remember being heartily laughed at by several of the officers . . . for quoting the Bible as an unanswerable authority on some point of morality."

FitzRoy might have made the voyage even without the Admiralty's backing—he had outfitted a ship and was ready to do so when they stepped in—because he had another important task: to return three captured indigenous people from Tierra del Fuego to their homeland.

Darwin's was the second voyage of the Beagle to South America. The first trip departed England in 1825 under the command of Captain Pringle Stokes. Like many other travelers, Stokes found Patagonia dreadfully dull. Faced with the idea of more time there, he shot himself. FitzRoy was his flag lieutenant, and he took over the Beagle and set off to survey Tierra del Fuego. Almost immediately, a group of Fuegians stole one of his whale-boats. FitzRoy gave chase but never recovered the whale-boat and instead took hostages, hoping that this would force its return. Much to his surprise, however, the Fuegians appeared satisfied with that bargain. He later traded another family some beads and buttons for a boy, and, with four indigenous people on board, FitzRoy seized on the idea that he would take them back to London with him, have them educated, and return them a few years later. He gave the three men and one woman the names York Minster, Boat Memory, James Button, and Fuegia Basket, and soon they were meeting the queen, ice-skating, and learning to eat with utensils.

Although Boat Memory died of an illness shortly after arriving in England in 1830, the others boarded the

Beagle with Darwin a year later and set off for South America, along with a missionary, to convert their friends back home. "Jemmy" Button was the "universal favourite," according to Darwin, and he remembered Jemmy sympathizing with his seasickness. "He used to come to me and say in a plaintive voice, 'Poor, poor fellow!'" Darwin wrote. "But the notion, after his aquatic life, of a man being sea-sick, was too ludicrous, and he was generally obliged to turn on one side to hide a smile or laugh." York Minster, the oldest of the group, was mostly quiet and reserved; Darwin described him as "taciturn" but "violently passionate" when excited. Fuegia Basket was the only female, and Darwin described her as a nice girl and a quick learner.

All the Fuegians taken by FitzRoy except York Minster were part of a tribe called the Yamana. The Yamana were semi-nomadic, living in small, self-contained family groups, and building small huts near oyster beds along the coast and then moving when the oyster supply gave out. They went back and forth between locations, returning to old huts after intervals long enough to allow the oyster stock to replenish. They also clubbed cormorants and seals, fished, and speared the occasional guanaco. They didn't wear clothing, and smeared themselves with seal grease to stay warm. They carried all their valuables with them in canoes, including torches brandished to help keep their fires alive, and when there was a big event, like a whale washing ashore or an English boat landing, they would all g a t her in their canoes, build their huts (it took only an hour or two), and hang out.

The Yamana communicated with each other by lighting signal fires along the coast—Magellan saw these in the 1500s and named the place *Tierra del Fuego* or "Land of Fire" in Spanish. The Fuegians lit a large signal fire when the Beagle arrived this second time. FitzRoy and Darwin observed it from the boat, and FitzRoy recorded being "astonished at the rapidity with which the Fuegians produce this effect . . . in their wet climate, where I have been, at times, more than two hours attempting to kindle a fire." Although the captain and his naturalist might have seen this as an example of the Fuegians' clever adaptation to their environment, neither did so. When they landed the next day, it was amongst "savages."

Darwin was transfixed. Landing amongst the Fuegians, he wrote, "was without exception the most curious and interesting spectacle I ever beheld. I would not have believed how entire the difference between savage and civilized man is."

The musical treated its Fuegian heroes with a similar level of condescension. Jemmy Button's actor turned him into a happy, effeminate Englishman who didn't want to leave England, wearing gloves and shiny pink clothing and sipping tea with a raised pinky. (Darwin did describe Jemmy as "vain of personal appearance.") The musical implied a romantic relationship between him and FitzRoy, which could have been Argentine actors subtly expressing their disdain for the English, or could have been overacting. Or it could have been entirely unintentional.

Whichever the case, the show proceeded to the point of separation, in which FitzRoy, having shepherded his Fuegians for the past several years, finally arrived back near their home.

On January 18, 1833, FitzRoy anchored at the eastern end of the Beagle Channel and set off with Darwin, a small crew, and the missionary, whose name was Richard Mathews, to return the Fuegians to Jemmy Button's old homeland in Ponsonby Sound, an inlet on the south end of the channel opposite modern Ushuaia. The yawl, a small sailboat used for exploratory expeditions, carried all of the items selected by Mathews' Missionary Society to help the newly converted turn from savagery to proper Victorians. Darwin noted with disgust that the outfit included wine glasses, tea-trays, fine white linen, and a mahogany dressing case.

When the Beagle crew turned into Ponsonby Sound after three days of sailing, they found a small cove identified by Jemmy Button as his former home, Wuluaia. They landed and on the next day, Jemmy's family arrived. (In the musical, the family was represented by dozens of grunting Yamana Indian puppets in canoes, which surrounded the stage.) Darwin remarked on the Fuegians astounding good eyesight and hearing: "all the organs of sense are highly perfected; sailors are well known for their good eyesight, & yet the Fuegians were as superior as another almost would be with a glass." The Beagle crew optimistically set to work clearing the land for European-style living, building houses for Mathews and his prospective neighbors, and planting vegetable gardens. On

January 28, FitzRoy decided that enough had been done and, leaving the three Fuegians and the missionary behind to let them get settled, turned the boats around to take a quick tour of the Beagle Channel.

As they sailed back toward Wuluaia a few days later, FitzRoy noticed a group of Fuegians, whom he didn't recognize, wearing white linens, which he did. Mathews was not the kind of missionary who would give away the shirt from his back. FitzRoy rushed onward.

Mathews, much to the captain's relief, came out to greet the ship, looking mostly as they had left him. But his report was unsatisfactory. He had had almost everything he owned stolen. He had been threatened. His garden had been trampled. "He did not think himself safe among such a set of utter savages as he found them to be, notwithstanding Jemmy's assurances to the contrary," FitzRoy wrote. The Richard Mathews experiment had finished, and on February 7, FitzRoy packed Mathews onto a whale boat and sent him back to the Beagle. At least Jemmy, Fuegia Basket, and York Minster seemed to have reintegrated nicely. Some vegetables were sprouting in the garden. FitzRoy felt sanguine.

The Beagle returned to Wuluaia one year later, in February 1834. They found the wigwams empty and no sign of Jemmy onshore. Soon a canoe appeared flying a flag. "Until she was close alongside," Darwin wrote, "we could not recognize poor Jemmy."

"It was quite painful to behold him," Darwin lamented. "Thin, pale & without a remnant of his clothes, excepting a bit of blanket round his waist: his hair, hanging over his shoulders; & so ashamed of himself he turned his back to the ship as the canoe approached. When he left us he was very fat, & so particular about his clothes, that he was always afraid of even dirtying his shoes; scarcely ever without gloves & his hair neatly cut. I never saw so complete and grievous a change."

FitzRoy rushed Jemmy below deck and had him clothed in English finery, and they proceeded to take tea together. Jemmy still remembered English and had even taught some to his family and to his new wife, who introduced herself to the sailors as "Jemmy Button's wife." The next morning, Jemmy told FitzRoy what had happened in the past year. York Minster and Fuegia Basket had stolen all of Jemmy's clothes and left in the middle of the night to return to York's homeland. And a warring tribe from the northeast of Tierra del Fuego, called "Ohens" by Jemmy—now called the Onas—had raided the settlement, forcing Jemmy to flee to his own island.

Still, Jemmy did not wish to return to England. He distributed some gifts, including a few spearheads for Darwin and a bow and quiver full of arrows for his schoolmaster in England. The gifts were as comically inappropriate as the tea trays sent the other direction by the missionary society.

Perhaps the musical had absorbed a bit of Darwin's grief. When the Jemmy actor reappeared near the end, he had lost the tight breeches and teacup and had acquired a club and a loincloth. His hair was longer and stringier. He presented FitzRoy with an animal skin and intimated via grunts that he didn't want to go back to England now, because he was happy being a savage. Jemmy and company (still played by puppets) then grunted their way offstage while a deeply chastened FitzRoy took the spotlight for his final solo, lamenting his pride and selfishness. "I played with this boy's soul," he crooned. "I took a life and treated it as if it was mine, my own to guide, to take to a better world."

The real FitzRoy, not quite so tormented, reluctantly sailed out of Ponsonby Sound after two days. "Every soul on board was as sorry to shake hands with poor Jemmy for the last time, as we were glad to have seen him," Darwin wrote. "I hope & have little doubt that he will be as happy as if he had never left his country; which is much more than I formerly thought." FitzRoy spoke optimistically of the small benefits that might be gained from reintroducing Jemmy, York Minster, and Fuegia Basket to their homeland. "Perhaps," he speculated, "a shipwrecked seaman may hereafter receive help and kind treatment from Jemmy Button's children."

In the 1845 edition of *The Voyage of the Beagle*, Darwin provided one last footnote about the Fuegians. The Beagle's lieutenant, Philip Sulivan, eventually became a captain and was assigned to survey the Falkland Islands in the 1840s. In 1842, Sulivan heard from a sealer that a woman had boarded his boat in the Straits of Magellan and that she spoke English. Darwin wrote, "without doubt this was Fuegia Basket. She lived (I fear the term probably bears a double interpretation) some days on board."

After the singing FitzRoy had belted out his musical lesson, the show reached its conclusion. The characters exited the stage and the video screen came down again, with the old version of Darwin saying it was too bad he and FitzRoy didn't see eye to eye on the whole evolution thing, but that the Beagle trip had been a good time all around. Then the sailors came back and belted out one more rousing chorus of "Our spirits will never die / The Beagle is flying high," and the stage went dark.

I tracked down Marcos Gomez, the blue-eyed actor who had played Darwin, after the show. I found him in the Beagle Center lobby, watching the cruise ship passengers enjoy a nice local seafood dinner under a poster of naked Fuegians. Gomez was twenty-five, with blond hair still damp from the performance, and as he peeled back his rubber sideburns to reveal a hidden microphone, he told me he had studied acting in Buenos Aires.

I asked if he studied Darwin.

"Yes, a little," he said. "To understand the part." Gomez paused and looked thoughtful. "He married his cousin," he added.

"Yes," I said. "That's right. And do you like playing Darwin?"

"Darwin is a fun character to play. But I would like to be the captain. He has a stronger role." He pantomimed singing and then shrugged. "You know. But last year, I was a sailor." He found his picture in the program and pointed it out.

I asked what he thought of the Yamana, and he became much more animated. "They're considered the most primitive of the tribes," he said. "But really, they were very smart. They had lots of means of obtaining food. They lived in complete harmony with the land. If a whale washed up, they'd eat that. They covered themselves in grease for the cold." He continued praising the Yamana, clearly enthused by the subject.

I wanted to ask how he felt about the musical portraying them as grunters, but Gomez had to run backstage to change. I lingered for a few minutes more then headed back to the hostel.

The next morning, looking for an academic of sorts to explain a bit more about current perceptions of Fuegians, I made my way through swirling mist to the local history museum. It called itself the "end of the world" museum, a tribute to Ushuaia's self-declared status as the southernmost city in the world. (Like most everything Argentina claims, this is disputed by Chile, which says that its own naval settlement, Port Williams, is further south.)

The Yamana merited only a brief mention alongside the names and dates of famous European explorers who had visited the area. "They had a number of beliefs commonly found in other primitive peoples," one plaque read (in English). "Yamana stories are known to be simple, with little humor or keenness." I went to the front desk and asked if I could speak to the director and soon found myself ushered into the office of a very professorial-looking Argentine. He lit a cigarette and introduced himself as Santiago Reyes. We chatted briefly about Darwin's visit. I asked what people in Tierra del Fuego, other than people who worked in tourism, thought of Darwin. "They don't," he answered quickly.

Reyes burned through three cigarettes and then switched to drinking *maté*.

"Darwin criticized slavery," I said. "He defended the Indians of Patagonia. Then he arrived in Tierra del Fuego and was very critical of the indigenous people. He called them savages. Why do you think that was?"

Reyes took a sip from the *maté* gourd and reached back into his desk to pull out a postcard that he then slapped on the desk in front of me. It was a black-and-white picture of a Yamana family squatting in front of some trees, naked except for loincloths. It looked just like the large wall mural in the lobby of the Beagle Center.

He pulled out another postcard, of another group of Indians from the northern part of the island. In this picture, the people wore coats made from guanaco furs and stood upright with good posture. "What do you see?" he asked.

"They look different," I said.

"There is a difference," he replied. "To a European, these Indians look better. Darwin was very young. With very little experience. The Indians of the north were more normal for him. Imagine what he saw here. The natives were dirty. They didn't have clothes. They're one-and-a-half meters tall."

Reyes didn't excuse Darwin's naiveté, but his side-by-side comparison provided an explanation. For a 23-year-old from a land where everyone wore wool and lived in brick houses and drove horse-drawn carriages, people sleeping naked in the rain must have seemed a breathtaking contrast. Except for the anthropologists who have discovered new Amazonian tribes, the experience Darwin had in meeting the Fuegians cannot be replicated today. Seeing rural farmers sleeping outside in sub-zero weather in the Andes or the people living in *favelas* in Rio de Janeiro may inspire a similar feeling of wonder at the differences between us, but I imagine it's nothing like the shock Darwin felt. This wasn't just the most curious and interesting spectacle he ever beheld, it was something he repeatedly referenced later in his life, even in his scientific work. Clearly, the Fuegians had set Darwin thinking: How are we the same and yet so different? Where do these vast differences come from?

The Descent of Man, published in 1871, isn't just Darwin grappling with a lifetime of abolitionism. It's his attempt to finally answer a question that had leaped out at him nearly four decades earlier in the savage conditions of Tierra del Fuego.

That night in the hostel kitchen I cooked an Argentine steak, sat down to eat, and found myself opposite a young traveler. He slouched in his chair, pushed his noodle dinner around with his fork, and looked out the window at the overcast night sky. We exchanged the standard introductory formalities, and after learning he was nineteen, from Canada, and named Thomas, I asked how long he had been in Ushuaia.

"Too long, really," he said. "This town just doesn't do it for me. It's too much like Canada. Maybe for you, coming from California, this weather is new or something. I can get rain and snow at home. They've got the same scenery. Same forests. They even took our beavers."

(Beavers—small, furry, cute beavers—have become the brown scourge of Tierra del Fuego. Someone had the idea of introducing them in the early 1900s, to be hunted for their pelts. The pelt market collapsed almost immediately, but the beavers liked the scenery and decided to stay on and eat the island alive. The Argentine government started offering a small bounty for beaver pelts, but it didn't help much. A local farmer explained to me that you could go out to the beaver pond, spend all day freezing cold for one decent shot at one beaver, and, if—*if*, he emphasized—you managed to hit one, all the other beavers within a ten-mile radius would vanish for the next few weeks. The farmer told me that they were considering introducing grizzly bears as a pest-control method.)

"At least," I countered, "your beavers have left a trail of destruction and carnage."

"That's true," Thomas said, nodding. "No natural predators. Reproducing like crazy. I saw all the dead trees in the park today."

Thomas was impressed by the beavers. They were, he said, probably the most interesting part of the whole Tierra del Fuego experience.

He looked over at the book I was reading.

"Charles Darwin," he said. "Why have I heard that name before?"

9: BAHIA BLANCA

The Gaucho Lifestyle

Upon reading my passport, & finding that I was a Naturalista, his respect & civility were as strong as his suspicions had been before. What a Naturalista is, neither he or his countrymen had any idea; but I am not sure that my title loses any of its value from this cause.

—BEAGLE DIARY, SEPTEMBER 19, 1832

THE NORTHERN INDIANS OF ARGENTINA, to Darwin's European eyes, weren't savages. But they were dangerous, especially for wandering collectors, because while Darwin was traveling through, Argentina was in the process of a genocidal campaign of eradication led by the capable General Juan Manuel de Rosas. The general was forty years old and had spent a lifetime accumulating land and power across the Argentine plains. "He will be a Catholic and a military man," his father told the chaplain at his baptism, and this prophecy proved eerily accurate—though not, likely, in the manner intended. Rosas expanded his ranching empire through a series of savvy business ventures, and he exercised rigid control over his workers—Argentine biographer Pacho O'Donnell suggests Rosas had a deep-seated fear of anarchy—whom he then recruited into militias. His courage and discipline inspired ferocious loyalty in his followers, and in 1829 he became governor of Buenos Aires. In 1833, no longer governor but still powerful, he set off to conquer the indigenous people of Argentina. The fighting was shockingly bloody and brutal, inspired on the Argentine side by a manifest destiny-like belief that the land was needed for white settlers and cattle ranches. The Indians, a roughly unified alliance of major tribes from across southern Argentina and Chile, fought a guerilla campaign out of the Andean foothills.

Darwin observed the ensuing wars firsthand. In August 1833, a few months after his first Tierra del Fuego trip, Darwin had the Beagle drop him off in Northern Patagonia, in a town called Patagones, where Rosas had secured a coastal strip that made it safe for Darwin to travel overland without fear of an Indian attack. Rosas himself had set up camp on the banks of the Colorado River, eighty-five miles north of Patagones. Darwin determined to visit the camp and then make his way overland to Buenos Aires. While the Beagle worked its way up the coast toward the port town of Bahia Blanca, Darwin recruited horses and a guide to take him north, into the pampas. To his good fortune, bad weather delayed the trip and allowed five gauchos to join the party. The benefit wasn't just protection. Darwin would get the full gaucho experience, and he would later be quite proud of himself for living the free-roaming, red meat and open country lifestyle.

On the first night, one of the gauchos spotted a wandering cow, and soon a barbecue was being prepared. "We here had the four necessaries for life 'en el campo'," Darwin wrote. "Pasture for the horses, water (only a muddy puddle)—meat—& fire wood. The gauchos were in high spirits at finding all these luxuries, & we soon set to work at the poor cow."

Darwin and his new friends shared the meat, spread their saddles on the ground, and went to sleep under the open sky—true cowboy style, and a mode of traveling that stirred something in the English naturalist. "There is high enjoyment in the independence of the Gaucho life," he recorded, "to be able at any moment to pull up your horse and say here we will pass the night. The death-like stillness of the plain, the dogs keeping watch, the gipsy-group of Gauchos making their beds around the fire, has left in my mind a strongly marked picture of this first night, which will not soon be forgotten." It was a sentiment he would express often in the next few years.

Two days later, Darwin arrived at General Rosas' encampment. On August 15, the great leader sent Darwin a message: He would be happy to meet the English naturalist.

"General Rosas is a man of an extraordinary character," Darwin noted in his journal. "He has at present a most predominant influence in this country & probably may end by being its ruler." In earlier editions of the published account, Darwin added, "which it seems he will use to its prosperity and advancement." A small footnote appears in the 1845 edition: "This prophecy has turned out entirely and miserably wrong."

Darwin needed Rosas' help in traveling through the country. He wanted a passport and permission to use government horses to travel between checkpoints established by Rosas—as part of his campaign, Rosas had established twelve armed outposts, or postas, between his camp on the Colorado River and Buenos Aires, and Darwin intended to follow the trail left by these small forts back to the capital. He didn't say much else about his conversation with Rosas, but he received both passport and permission "in a most obliging manner."

"My interview passed away without a smile," Darwin dryly concluded, declaring himself "altogether pleased with my interview with the terrible General. He is worth seeing, as being decidedly the most prominent character in S. America."

It later turned out that even the fact of the meeting itself was useful. At the end of his long ride Darwin tried to enter the city of Buenos Aires only to find it blockaded by revolutionaries allied with Rosas. He was barred from traveling overland, and there were embargoes on all the ports—until Darwin mentioned his meeting with Rosas. Instantly, the leaders gave him permission to pass.

"Magic," Darwin wrote, "could not have altered circumstances any quicker."

My interest was in culture more than politics. Rosas was the unofficial king of the gauchos, and Darwin had met with probably the world's most powerful cowboy until Ronald Reagan. The general was famous for his feats of horsemanship, his traditional dress—he once called on the British minister to Buenos Aires in full gaucho regalia—and his reputation as a practical, no-nonsense man of action. With his downfall and eventual exile to England in the mid-1850s, gaucho culture lost its most visible embodiment. By the end of the century, as private landowners increasingly forced the gauchos to abandon their wandering lifestyle and settle on the ranch, the culture of plains-roaming, free-spirited, egalitarian nomads had all but vanished.

I wondered what that meant now. The gaucho endures in Argentina and Uruguay as a kind of tourist icon, something found on postcards in Buenos Aires. For reasons I never quite fathomed, it's also the athletic mascot of my undergraduate alma mater, the University of California Santa Barbara. I was curious to find whether anything beyond that had actually survived. Was there anything to the gauchos today, except for a pleasing set of values, a historical idea, and a Zorro-masked mascot?

I took a bus to Bahia Blanca, the next town north of Rosas' post on the Colorado River, and arrived at 3 A.M. I slept four hours on a bench in the bus terminal and then staggered downtown to see a city that had grown by leaps and bounds since Darwin visited and reported, "Bahia Blanca scarcely deserves the name of a village." It was afternoon by the time I made it to the heart of the city, a leafy plaza surrounded by cafés and shops. In the middle of the plaza, old men sat around stone tables and played cards while families picnicked and young couples groped each other in the shade. Wandering down one street, I came to *El Matrero*—a veritable gaucho emporium. Leather saddles and halters, colorful ponchos, and paintings of romantic gauchos jammed the dark walls. Display cases full of *maté* gourds and silver knives took up all the space in the middle of the store. Everywhere I stood I felt in the way, bustled and pushed by busy store clerks searching for gaucho relics. The jostling pushed me toward the back of the room, where a wrinkled, silver-haired leather smith stood, smelling of leather and glue and working on a worn saddle blanket. He smiled as I approached.

"You put this under the saddle," he explained in Spanish, holding up the edge of the blanket, "and tie a belt around the horse here. The buckle on the belt rubbed against this and broke the stitching." He pointed out where the yellow stitches had snapped and poked a four-inch needle into the blanket.

I asked the man working the saddle what he thought. Was gaucho culture dead?

"Gaucho culture dead?" he said. "Noooo. In the countryside there still are gauchos, certainly. They have changed, of course. But they are still gauchos. Certainly."

The array of gear behind the man looked a bit like the kind of stuff you'd find in one of the more interesting stores in San Francisco—lots of jangly, studded, leather. A rack of *boleadores*, one of the traditional hunting weapons of both the gauchos and the plains Indians, hung over his head. The *bolas* consist of two heavy, metal spheres or rounded rocks, about the size of a baseball, connected by a long leather strap. Hunters whirl the balls around their head and then throw them at the legs of animals, entangling the animal and tripping it so it can be finished off with a spear or knife. Darwin tried his hand at both the *lazo* (lasso) and the *bolas*, with less-than-perfect results. After riding hard, whirling the *bolas* around his head, he tossed them—and snared his own horse. The gauchos, he reported, roared with laughter. They'd seen all kinds of animals caught with the *bolas*, they teased Darwin, but they'd never seen a rider catch himself.

I asked the man in the saddle shop if he could use the *bolas*.

"Me?" he said, surprised at the question. "No. The gauchos use them. Not me."

I picked one up, briefly, and was surprised at the weight of the metal ball. I put it back. The leather smith smiled and set aside his saddle to sip his *maté*.

In the front of the store, a flier advertised a performance of gaucho *folklórico* dances that evening in the community center on the edge of town. A few hours later, a taxi driver dropped me on a small, dark street and pointed back at an unmarked building. The inside reminded me of my high school gymnasium. Concrete bleachers cascaded onto a smooth concrete floor, and folding tables and chairs clustered around the edges. A group of teenage girls operated a concession stand in the corner. I bought a soda from them, and one of the girls caught my accent. She asked where I was from. "And you came here for this?" she asked, incredulous.

The dancing started around 10 P.M. with a crowd-pleasing dance by the youngest troupe in attendance. Dressed in oversized gaucho duds or Spanish ranch finery, ten children tripped out under the lights to perform a simple partner dance. The boys wore huge baggy pants, knee-high leather boots, white long-sleeved shirts, and red kerchiefs and sashes. They covered up with tightly cut red, black, or brown jackets, like Spanish matadors. Most also donned the low, flat, black sombreros fancied by the gauchos. The girls, meanwhile, wore ornate long dresses and pulled their hair back and up. For the dance, both boys and girls raised their arms and moved around each other in stumbling circles, vaguely in time with the music and under direction of their coach, who stood alongside, waving to remind them of positions. One boy lost his red sash, which he kicked aside with a smile. Another boy and girl forgot their steps and excused themselves to stand and wait for everyone else to finish. The audience of several hundred *maté*-drinking family members and friends loved it. The kids trickled back out of the lights to enthusiastic applause.

I went in search of someone who could explain the dances that followed and found Paula Gil, a twenty-eight-year-old professional dance instructor from Bahia Blanca.

She pointed me up to the top of the bleachers, where her girls-only troupe prepared for its performance later in the evening. Watching from above, we could see dancing couples circling each other. Every once in a while, the boy would stop dancing, turn his back to the girl, and start to clap in time with the music. The girl sashayed around him, working closer and closer, until she ended up right in front of him. They would split apart again and continue dancing.

The dances, Paula told me, represented a specific period in Argentine history, from about 1815 to 1870, and resulted from a blending of Spanish and Italian culture.

"Are many people interested in *folklórico*?" I asked.

"Of course more people do the tango," she said. "It's more erotic, more difficult. But this is the dance of the countryside."

That reminded me of what the man in the gaucho store had said. "I've seen a few books that say that gaucho culture is dead," I said.

"There are huge cities now," she replied. "Buenos Aires, Córdoba, Mendoza. In the cities, yes. But in the villages, outside the city, the culture still remains important. The customs are entirely from the gauchos. It's a time period that's a part of our history, so people want to learn."

Below us the dancers split apart, and the men started clapping. "This is an *escondido*," Paula said, switching subjects to explain the dance. "One dancer can't see the other. The other is hidden. So to help the other dancer find him, he makes noise." The women circled in on their clapping partners, and everyone started dancing again.

Paula's students performed later on. When they came back to sit down, I asked her about their stomping, clapping dance. "It was an *escondido*, right?"

"You're learning!" she said, looking pleased. "You have to keep teaching these dances," she said, looking on as the next group started. "They were forgotten through the last century, but people started to learn them and teach them again. Now we have to teach each generation."

On stage, yet another group of kids negotiated the tricky footwork of the gaucho dance. The clock struck midnight, and I looked out over the crowd, still very much jumping. It looked like it might turn into a typical Argentine social occasion—which meant it would probably go until 5 A.M. I hadn't slept since the four hours in the bus terminal, so I caught a ride back downtown from a pharmacist who had to get back to work. As we left, the live music had just started.

The next day I took a bus into the more rural mountains outside Bahia Blanca, following Darwin's overland route. As I traveled, I pondered the gaucho dance, and I thought about Paula telling me that it was important to keep the old ways remembered. Argentines, at least in places like Bahia Blanca, seemed to be doing that for the gauchos. But what about the Indians? As I arrived in the small resort town of Sierra de la Ventana, I walked out to see if I could find out more about the people lamented by the Port San Julian historian, Pablo Walker, and others, as mostly forgotten.

There wasn't much information out there. In Darwin's time, the significant tribes in Patagonia included the Mapuche and Tehuelche. But the Mapuche were linked mostly with southern Chile, where they'd resisted colonization not just by the Chileans and Spanish, but by the Incas as well. Unconquered into the late 1800s, the Mapuche eventually were forced into a reservation system that still exists in Chile. The Tehuelche, meanwhile—the people encountered in Port San Julian by Magellan—were associated more with southern Argentine Patagonia, where a few thousand still live. These were the tall, powerful Indians generally mentioned by European explorers. But the northern Tehuelche, who differed linguistically from the southern groups, disappeared by the end of the nineteenth century, killed by the Spanish or absorbed into Mapuche societies.

Darwin, who encountered both northern and southern Tehuelche, had predicted a grim future for all the Indians, writing in his diary that at the rate Rosas's extermination campaign was carried out, "in another half century I think there will not be a wild Indian in the Pampas North of the Rio Negro. The warfare is too bloody to last."

The Indian wars didn't act on Darwin's mind the way slavery did, and he wasn't a champion of indigenous rights. For one thing, although he sympathized with them, he still feared an attack. Almost all fireside conversations, he reported, eventually ended up in discussions of Indian raids. He had two trips outside of Bahia Blanca interrupted by scares, although both turned out to be false alarms.

When he got a chance to study friendly Indians, he was fascinated. "My chief amusement," he wrote from one military camp, "was watching the Indian families as they came to buy little articles at the Rancho where I staid." Rosas had numerous Indian allies, and so there were plenty of people for Darwin to observe. "The men are a tall exceedingly fine race," he wrote approvingly. "Amongst the young women, or Chinas, some deserve to be called even beautiful."

I asked for town history at the tourism information office in Sierra de la Ventana, and the woman handed me a thick three-ring binder. "This is all the history," she said. The folder included "recent histories" of all the nearby towns, city maps, ecological studies, and park histories, even a list of references and a course layout from the city's first country club. "All the history" didn't include much about the Indians though. One document mentioned that the fertility of the valleys and protection of the mountains made Sierra de la Ventana an appealing place to shelter from Rosas's raids, and also one great battle in the extermination war, on the plain near the town of Tornquist, where two hundred Indians were killed. Darwin commented on the battle in his own journal: "The Sierra de la Ventana was formerly a great place of resort for the Indians; three or four years ago there was much fighting there; my guide was present when many men were killed; the women escaped to the saddle back & fought most desperately with big stones; many of them thus saved themselves."

When I had finished with the binder I returned it to the office and tried to find the town library. In a building next to the school, a middle-aged, slightly roundish woman with short black hair and glasses sat surrounded by translated bestsellers by Dan Brown, Michael Crichton, John Grisham, Stephen King, and Danielle Steel. We were interrupted every few minutes by schoolchildren in white lab coats needing help finding books.

Sara Bellabarba became animated when I asked about Indians. "It's very difficult to tell the history of this place because we don't have it," she said. "There is very little about the indigenous people here. And about the war, it's also difficult. I don't agree with it, because we moved onto their land and they fought back. If someone invaded my house, there would be a battle."

Darwin wasn't inspired to act, but he despised the cruelty of the Indian wars. One soldier reported to him the "unquestionable fact, that all the women who appear above twenty years old, are massacred in cold blood." Darwin, shocked, called such murders inhuman. "He answered me, 'Why what can be done, they breed so.'" In other conversations, Darwin noticed that everyone believed the war just, "because it is against Barbarians." He concluded in his journal, "Who would believe in this age in a Christian, civilized country that such atrocities were committed?"

10: SIERRA DE LA VENTANA

Cerro Tres Picos

> *I had now been several days without tasting anything except meat & drinking mattee. I found this new regimen agreed very well with me, but at the same time felt hard exercise was necessary to make it do so.*
>
> —BEAGLE DIARY, SEPTEMBER 17, 1833

SIERRA DE LA VENTANA has roughly one ice cream store for every four hundred residents. Darwin's primary purpose in visiting this sleepy place—there wasn't a town or an ice creamery at the time—was to try to climb a nearby mountain. An inveterate hill-hopper, Darwin almost always found joy in climbing. But there was little joy in this expedition. "I do not think Nature ever made a more solitary desolate looking mountain," he wrote from the approach. "It is very steep, rough & broken. It is so completely destitute of all trees, that we were unable to find even a stick to stretch out the meat for roasting, our fire being made of dry thistle stalks." The next day he became the first European to explore the highest mountain in the modern-day province of Buenos Aires, the 4,060-foot-tall Cerro Tres Picos.

The mountain shot up over the town from a large private estancia, and in the evening I called the ranch's tourism coordinator and asked whether I could hike it. "Come early," she said. "It will take you ten hours. If you don't leave by nine, you can't climb the peak."

Darwin's guide gave him faulty information, telling him to ascend a nearby ridge that he could follow to the summit. But when Darwin reached the ridge top, he found a deep valley separating him from the highest peaks. He had to scramble down and then up the other side. By the time he reached one of the minor peaks the day was late and he was suffering from cramps in his thighs. He had to give up the highest point, and his frustration spilled out into his diary: "Altogether I was much disappointed in this mountain; we had heard of caves, of forests, of beds of coal, of silver & gold &c &c, instead of all this, we have a desert mountain of pure quartz rock."

I had spent enough time following Darwin around, reading his journals, and trying to channel his thoughts, that I felt almost certain he was disappointed with the scenery mainly because he hadn't reached the top. I knew myself; faced with mountain failure, I'd be frustrated, and my journal would reflect that. And I knew Darwin well enough to recognize that, at least in the matter of big bad-mountain climbing, we shared a certain summit-or-bust aesthetic. This would also be my last real travel day on the east coast before I caught a bus back to Buenos Aires to head home, making a good concluding mountain vista imperative.

I determined to reach the highest peak and take in the view. The next morning I hired a cab to drive me to the Estancia Funke, the private ranch that encompasses the range of mountains where Darwin explored. The cab dropped me at sunrise near a cluster of tall trees growing by the back porch of one of several ranch houses. A sign identified it as the tourism office, and I knocked on the screen door. Monica Silva, whom I had spoken with the night before, came out to greet me, accompanied by a porcine black lab that promptly buried its slobbery head in my crotch. I patted it tentatively and it wagged its tail and sat at my feet, drooling on my shoes. Silva looked ready to go for a hike, in nylon cargo pants, fleece jacket, and hiking boots, and she was much younger than I had imagined from her hoarse phone voice. She was tall and tanned, with dirty blond hair and freckles. I saw a card on the door introducing her as Professor Monica Silva. "Physical education," she said, when I asked if she studied history or geology. She handed me a book of liability forms and a deposit form in case I needed rescue, and while I signed away my life, told me briefly about the ranch.

Rodolfo Funke came to Argentina from Germany in 1877 at age twenty-five, fleeing political and social problems, and met a friend from home named Ernesto Tornquist, who had established himself as a prominent landowner. Tornquist was also a friend of one of the top Argentine generals. A year after arriving, Funke took a solo trip on horseback through the mountains of the Sierra de la Ventana and, unlike Darwin, liked what he saw. Tornquist offered to sell him the land, and Funke bought 3,700 acres so he could start a ranch. In 1940, two years after Funke's death, the newly founded Hogar Funke Foundation began to transform the ranch by converting seventy-five acres, including the mountain peaks, into a park.

Without expecting much, I asked Silva about the Indians who had lived in the area before Tornquist and Funke divided up the land, and I got the familiar answer. "Putting it all together is very difficult."

No other tourists had called at the ranch, and when I finished the forms, she explained the climb. "It's about ten hours," she said. On a rough map, it looked to be about five to six miles and a 3,000-foot climb to the summit. Counting on her fingers, she added, "You should turn back at one o'clock." Then she gave directions.

"First," she said, "you're going to hike five kilometers to the Glorieta Outpost. You'll pass two cattle guards on the road, then a yellow house on your right, then go left. You'll cross a bridge and pass a few more cattle guards, and then you'll see the sign for the outpost and a gate." She held up a picture of a small, circular sign on a post, and another of the gate and a cluster of trees around a small green bathroom. I nodded.

"Once you pass the gate, cross the creek. You only cross the creek once! From then on, always keep the creek on your right!" She waited to make sure I understood.

"Only cross the creek once. OK."

"Good. Keep the creek on your right and you'll come to a eucalyptus forest. Walk through the forest, and when you get past it you'll come to a gate. That's where the trail officially begins. From there, you'll be able to see a tree with a broken top on the ridge in the distance. Climb toward it."

She held up a picture of the ridge, with the tree with the broken top circled. It seemed awfully small. I wondered what would happen if the tree's broken top ever fell off or filled in with leaves again.

"When you get to the tree, turn and climb up the ridge," Silva continued. I tried to concentrate and found myself jumbling the directions together. "You'll see a corral at the top of the ridge." She held up a picture of the corral, a small wooden structure buried in high grass. "Just make sure you keep your back to the corral and hike up to the top of the ridge—but don't go over it. If you go over it, you'll go to the wrong peak." She showed a picture of the wrong peak, which I suspected was the one Darwin reached.

"Is that where Darwin made it to?" I asked.

"Yes," she said. "This one on the left here." She pointed it out on the picture.

"Anyway, when you near the ridge, you'll see another tree in the distance. Walk toward that and keep it on your left. Once you reach that, you'll see the peak. From there, you can climb it. Got it?"

"Sure," I said. She handed me a cartoon map, labeled "not to scale," with drawings of the creek, eucalyptus forest, broken tree, corral, and other tree.

"Good luck!" she said.

I started walking under the blue skies of a gorgeous early fall morning. Birds twittered in the trees. Butterflies fluttered across the road like falling leaves, blowing in the wind, sunlight turning them translucent. It took about an hour to reach the Glorieta Outpost, where I leaped over the very important creek and scrambled through a field of waist-high grass and star thistles. At the far end of the meadow, the trail wound up into the eucalyptus forest, rising sharply on steps of slippery white quartz and loose pebbles. At the top of the forest I came to the gate marking the start of the trail, a small wire fence amid fields of gently waving green grass. Slabs of rough gray rock punctured the ground in regular geometric patterns, making ridges resembling the backs of armor-plated stegosaurus. The path continued along the ridge toward the tree with the broken top, disappeared over a hilltop, and reappeared where the next hill rose in the background.

I passed the tree with the broken top and reached the corral, where a delicate breeze brushed the air clean of noise. I was well-along on the trail now, feeling much more certain that I would actually be able to carry out Silva's instructions (after some dark moments down below), and feeling not even remotely tired. The possibility of beating Darwin to the top gave me an extra boost of adrenaline.

At the top of the next ridge a small herd of guanacos eyed me curiously but without apparent concern and allowed me to get within fifty feet before slowly, grudgingly ceding the ridgeline. Darwin reported that if a sportsman suddenly encountered several guanacos standing together, "they will generally stand motionless and intently gaze at him; then perhaps move on a few yards, turn round, and look again." I remembered Darwin also describing them as inquisitive animals. "That they are curious is certain," he wrote. "For if a person lies on the ground, and plays strange antics, such as throwing up his feet in the air, they will almost always approach by degrees to reconnoiter him." I enjoyed the image of the well-dressed Englishman lying on his back, kicking his feet in the air, and rolling around on the ground in the middle of nowhere, so I decided to try it out myself. The guanacos hadn't gone very far, and readily allowed me close again. This time, as they started to walk away, I dropped my backpack and fell to the ground on my back, landing directly on a sharp, jagged piece of quartz rock. Fortunately the stabbing pain in my back added to my contortions, and the guanacos stopped walking, looked at me with heads cocked, but didn't approach. They may have rolled their eyes. I stood up, brushed the grass and dirt off and addressed the herd. "Thanks. You guys have been great."

The shattered rock summit rose like a giant serrated tooth, just beyond the guanacos. FitzRoy had spotted this peak from the harbor in Bahia Blanca, fifty miles away, inspiring Darwin to attempt to climb it. In his journal, Darwin described the highest part of the range as having four peaks in descending order, but I found only three—which made sense, since Cerro Tres Picos means three peaks mountain. "There are so many rocks sticking up here," Silva later told me. "Who knows what he saw."

A beaten-down guanaco trail led to the last ascent and then turned up into tumbled rock piles. Using my hands and knees I scaled the steep final portion and reached the top of the peak. Far below, green-brown and olive fields stretched out in a quilt across the land. The fields likely wouldn't have been there in the 1830s, but I was confident that frustration had clouded Darwin's opinion of the scenery. "I had hoped the view would at least have been imposing," he wrote. "It was nothing; the plain was like the ocean without its beautiful colour or defined horizon."

I felt like I had just one-upped Darwin (who was, by all accounts, exceedingly fit). I also felt like I'd had, for the first time, a full day that was approaching one of his full days. No people for the last several hours, and no guide. No trail to follow, just vague directions involving broken trees and creeks. A long, challenging hike with twelve miles of hiking and plenty of opportunities for teasing guanacos and studying geology. No giant city sprawled beneath me. But rather than wallow in the glory, I wanted to challenge myself further. I'd taken a taxi to the farmhouse that morning, but there was a bus stop at the highway six miles from the ranch. If I walked those six miles, rather than take a taxi, I'd have an even more Darwin-like day of exercise.

This is why, I think, I should not be allowed to make decisions from mountain peaks.

I skipped the six miles down the mountain and arrived back at the ranch house after just under seven hours of hiking. "Wow," Silva cheered when I knocked on the door. "Did you reach the top?" She asked me about my hike, interjecting *barbaro!*—or "cool!"—every three or four words. "*Que velocidad! Barbaro! Que velocidad! Barbaro! Y alcancaste el pico? Barbaro! Que velocidad!*" She asked if she should call me a taxi. I shrugged it off.

"No thanks," I said. "I think I'll walk."

Silva looked dubious. I insisted. "I'm not very tired," I said.

I hiked away from the ranch, past beautiful farmhouses and open, golden fields. Cerro Tres Picos grew gradually smaller behind me, fading from dark gray to light brown to a hazy purple, shrinking in comparison to closer peaks until it had disappeared. After two hours I reached the highway and sat down to watch the infrequent cars pass by. The sun started to set and the sky turned pink. A fox loped through the knee-high grass, crossed the road, and disappeared into some bushes. I felt an overwhelming serenity bubbling up within me. The last bus of the day came by at 6:45, and with a triumphant smile I held out my arm and took a few steps toward the road. The bus accelerated right past me. All of a sudden it was dark, and I was nine miles from the nearest city and twenty-one miles from my hotel, with eighteen miles and 3,000 feet of climbing already logged. I no longer felt serene.

I thought about hitchhiking, but the cars became even more infrequent. At first a car heading toward Sierra de la Ventana would pass every five minutes, and then it became every ten minutes, and soon, every twenty minutes. Rather aimlessly trudging along the side of the road, just to feel I was doing something, I wondered whether the cars could even see me. Eventually I spotted a sign for a small resort hotel and conference center two miles away, and I set out walking again, tripping over sticks and debris in the dark. I reached the hotel and woke the owners, who called me a taxi. "You walked from Cerro Tres Picos?" they said. "To here? Why?"

Getting trapped in the middle of nowhere was a risk Darwin constantly faced. In one worst-case scenario, he and seventeen crewmen got stuck on a beach within sight of the Beagle, unable to return to the ship because a squall whipped the water too much for a small boat to collect them. Darwin and his companions huddled together on the exposed ground to try and keep warm, and soon ran out of food. They ate seagulls and a washed-up hawk found on the beach and settled in to wait out the storm. "I never knew how painful cold could be," Darwin wrote after spending the entire night shivering. The next day the wind finally slacked, and everyone returned to the ship.

Coming down from his frustrating day on Cerro Tres Picos, Darwin at least had a camp to return to. And after "drinking much mattee & smoking several little cigaritos," he rolled out his sleeping bag and began to think a little more positively. He knew, after all, that though he hadn't reached the summit, he was the first European to explore the mountain range. Accomplishment always made him feel peaceful. Drowsy from the exercise, he dropped off to sleep immediately, and later reported, "It blew furiously, but I never passed a more comfortable night."

Determined to do the same, I lay down in the grass to nap while the taxi came to find me.

PART III
DISCOVERY

FOLLOWING DARWIN ON THE WEST COAST

11: CHILOÉ

Charming Green Things That Don't Ooze

In these shaded paths, it is absolutely necessary to make the whole road of logs of trees, such as described on the main road to Castro. Otherwise the ground is so damp from the suns rays never penetrating the evergreen foliage that n e ither man nor horse would be able to pass along

—BEAGLE DIARY, NOVEMBER 24, 1834

THE DAY AFTER CLIMBING CERRO TRES PICOS, I returned to Buenos Aires and from there flew back to California. My plan was to take a break from traveling, propose to my girlfriend, and then start researching the west coast of South America. In late spring, as I brushed up on Darwin's travels, I found myself talking to an old school friend. Josh Braun was trapped in Manhattan at a high-hour, low-pay magazine job under the direction of your industry standard brilliant-but-mad editor-in-chief. When I left for Chile a few months later, Josh was on the plane with me. He'd packed a bag full of radio equipment and talked about his plans to record stories about South American monster legends. I was still trying to figure out how it had gotten to this point. Although I welcomed the company—my travels on the east coast had often been lonely—I worried about how well Josh would be able to follow my Darwin enthusiasm. At least he was there for all the right reasons: Fleeing a bashing-the-head-into-the-wall post-college job, looking to see some of the world before he committed to something serious like graduate school or marriage.

Josh was tall, gangly, and distinctively red-headed—a look that instantly proclaimed irredeemable gringo-ness. I had always escaped gringo alarm bells because I'm generally swarthy and have a decent Spanish accent. Five-foot-eight with brown skin and brown hair meant I could easily pass for Brazilian, and I enjoyed the obviously conflicted touts who looked me up and down and then tried catching my attention in Portuguese. With Josh there would be no sneaking about.

Still, standing out might not be bad. Darwin certainly did. Josh was also deeply scientifically curious, and he had never been to South America before. He read voraciously and could speak generally on almost any field of science, though his recent studies had been largely confined to medicine and bioethics. He remained one of the few non-

NASA employees in the United States who passionately cared about the space program. It would be interesting to see how Darwinesque his response to the Chilean landscape would be. Plus, his father was a geologist, and so Josh knew a bit about rocks—and he was an enthusiastic hiker who wanted to decompress after a year in New York City. "I need to see green things that don't ooze," he told me, "and tall things that aren't buildings."

Southern Chile, with its abundant verdure, seemed to fit the bill. We decided to start in Chiloé, a rugged archipelago off the southern coast that someone had dubbed the "Ireland of South America." Guidebooks and tourist literature also called it "The Magic Isle." Darwin and the Beagle had visited after passing through Tierra del Fuego for the last time and proceeding north up the west coast. Chiloé was not their intended destination—FitzRoy wanted to skip most of Chile and head to the northern desert—but a series of gales forced him into port. "On leaving Tierra del Fuego," Darwin noted wryly, "we congratulated ourselves too soon in having escaped the usual course of its storms."

Chile is a remarkably thin country, with one major international airport in Santiago, tucked between the Andes and the coastal mountains. We plotted to leave quickly for Chiloé soon after landing; although it's clean and safe, Santiago is one of the least-interesting cities in South America, lacking the culture and food of Buenos Aires and the frenetic energy and distinct geology of Rio de Janeiro. "Of the town itself there is little to be said," Darwin wrote in his journal.

On our first day, as we adjusted to the jet lag and the rapid-fire syllable-dropping Chilean Spanish, I prepped Josh for our trip to the south. After forcing him to try a pisco sour, the national drink made from pisco and raw egg, and dragging him up the highest hill in Santiago, I pronounced him fit for Darwin travel.

At 10 P.M. the next night, we waited in the pouring rain at the Santiago bus terminal. Overnight buses in South America resemble giant insects, with antennae-like mirrors hanging over lit headlight eyes and smiling grilles, and as our bus pulled up, rain swirling in every direction, I felt we'd entered a hive in some prehistoric swamp, a swamp from the time sixty-foot wasps ruled the earth. We boarded without incident but when we woke up the next morning, somewhere outside the town of Puerto Montt, the bus television was showing scenes of traffic carnage and severe flooding in the middle regions of Chile. Residents fleeing their houses in boats, cars stalled as water flowed around their bumpers, yellow-suited men in waders riding around in helicopters.

Where we were headed, we imagined, had to be worse. "I do not suppose any part of the world is so rainy," Darwin wrote, "as the island of Chiloé." He complained in a letter home that "Chiloé, from its climate is a miserable hole." FitzRoy called surveying the archipelago just south of Chiloé "the survey of another Tierra del Fuego, a place swampy with rain, tormented by storms." The ship's Lieutenant Sulivan, less bothered or less literary than the other two, simply wrote in a letter, "It rained every day but one for six weeks, and most of the days never ceased raining."

Early in the morning we arrived at the ferry that would take our bus across to the island. The weather had cleared up and stabilized. It was a beautiful ferry ride—half an hour across placid gray water—and we stood on the deck and looked out at the island's dark green hills, dotted with white specks of houses, all of it delightfully pastoral and tranquil. So far, Josh seemed to be handling the culture shock well. He claimed to enjoy the food—hot dogs were the national dish—and other than a habit of smiling and blinking every time he tried to say thank you to someone, he was doing OK with the Spanish-for-n o n -Spanish-speakers. And as long as one of us spoke moderately well, most people reacted politely to our gringoness. Plus, Josh was genuinely excited about Chiloé. The island had a deep, fascinating mythology that he hoped to explore for his radio story. Best of all, when it rained here, it wasn't somebody tossing out the contents of a bucket from an apartment two floors up.

When Darwin arrived in June—winter, the wettest time of year—the weather had been perfectly benign for three days, and he was as surprised as we were. He recalled "the inhabitants themselves wondering at such an event." I asked the first person I could find, a shopkeeper in a small knick-knack store, about the unusual weather. "I had read," I said, "that Chiloé was always raining."

"Oh, it is," he said, matter-of-factly. "Lucky right now. But it'll rain tomorrow. Or tonight." He thought for a minute. "Or," he added, "both."

The bus dropped us in Ancud, in Darwin's time the capital of Chiloé. Smoke rising from chimneys and metal stovepipes filled the air with the smell of burning wood and left a low lingering haze over the town. "Darwin wrote that the first dry day he was here, the 'whole of Chiloé' was out burning brush to clear land," I recited to Josh as we walked downtown. Josh wrinkled his nose at the smoke and looked off into the shimmering, hazy distance, where there were supposed to be mountains visible on the mainland. "Well," he said hesitantly, "I suppose that's a nice parallel."

The town boasted a small grid of streets and a ping-pong and pool hall at the top of the hill. "I don't remember Jim Croce singing about ping-pong," Josh joked as we walked by a group of slouching, cigarette-smoking teens loitering in front of the building. This was typical Josh: Obscure, super-smart jokes that referenced songs that no one had listened to in thirty years. We kept walking. At the northernmost waterfront edge of the town, we strolled through the Plaza de Armas, a small leafy square with statues of different creatures of Chilotan mythology. We wandered up a small hill and found a coffee shop called "The Enchantment of Chiloé," where we ordered real coffee (a rarity in Nescafé-mad Chile) and watched as a pretty young waitress sashayed by while singing along, in stilted English, with the Annie Lennox and Aretha Franklin duet "Sisters are Doin' it for Themselves."

The "Enchantment of Chiloé" is a tourist draw: The

island's complex cast of folkloric characters had grown into a legend and spawned the "Magic Isle" moniker. But I wondered about it the same way I wondered about gaucho culture in Argentina. Was it something real? What was its historical basis? If twenty-something waitresses in coffee shops were serving espresso and watching VH1, how well had this leprechaun-and-Lucky-Charms-style mythology stuck?

"They are all Christians," Darwin wrote of the Chilotans, "but it is said that they yet retain some strange superstitious ceremonies, and that they pretend to hold communication with the devil in certain caves." He couldn't find out more, though, because, he reported, anyone convicted of such worship was sent to the inquisition in Lima. Now, the particulars are quite in the open. The island's belief system and tourism industry now proudly announce the virtues of Trauco, a troll-like creature whose ugliness makes him irresistible to women, and Fiura, his female counterpart (and, according to some versions, daughter). Fiura was supposedly just as ugly, but she was reported to have found a novel use for her bad breath; one huff and she'd overpower and sleep with any man she desired.

Curious to learn more about the mythology, and eager to find out whether people still believed it or just knew that it was good for scoring money off gullible tourists, we sought out the director of the regional history museum. Dutch-born archaeologist Marijke van Meurs referred us to Don Carlos Aguilar Cardenas, whose primary recommendation appeared to be that, unlike her, he was very old and had lived on the island most of his life. We found him out behind the museum, in a small stone tower with castle battlements, carving wooden ships with a chisel and chatting with a friend. Don Carlos—which, incidentally, is what Charles Darwin is often called in Spanish-speaking South America—was seventy-eight years old, gnarled and wrinkled, with one yellowed tooth poking out at a jaunty angle from his lower gum. His middle-aged friend, who introduced himself as Antonio Gia, seemed equally pleased to sit down and talk mythology with us.

We were lucky. The mythology doesn't appear to have any defining textbook, meaning that the legend of characters like Trauco is related in stories. And we had our own oral historian. If only, I thought as he started up, he didn't speak with such an absolutely indecipherable accent. A combination of every bad habit of Chilean Spanish and his own ruggedly shaped mouth meant that something like a third of his words never emerged into the world, and those that did came out in a kind of croaking blur, minus many of their most important consonants.

He emphasized Trauco, his favorite character. Short and squat, Trauco lived in forest caves and wore clothing made of dried straw. He was small, Don Carlos said, "but he had the strength of a Hercules." A tourist had once come in and seen Don Carlos working on a block of wood with an axe and shrieked, "Trauco!" His face lit up at the memory. "The axe, you know, is also part of Trauco's appearance," Don Carlos explained.

Trauco's mischievous trick was to impregnate single women. "Not so much now," Antonio told us, "but twenty or thirty years ago, in the countryside, if a single woman got pregnant, they said Trauco did it."

"Always, in jokes, we say that Trauco did it," Don Carlos added.

"And what do people think now?"

Antonio answered, while Don Carlos nodded. "What's happening now is that in the countryside, it's almost equal to the town," he said. "They used to be more reserved, that kind of thing. Now, everyone has a TV. Now everyone has a cell phone. Some have Internet. They're losing the culture as a result." He gave us an example: Even in the countryside, kids no longer listened to folkloric music. "They're all listening to rock music," he lamented. I thought back to our waitress in the coffee shop.

Don Carlos agreed. "Before, the people in the countryside, the ceremonies, everything was ceremonial and respectful," he said. "When you visited a place in the countryside, they attended to you like a king, like a prince." He paused for a minute. "Today, no. Today, nothing distinguishes the countryside from the city."

While mourning their lost traditions, Don Carlos and Antonio were also essentially describing the rural Chiloé that Darwin visited. An 1832 census counted 42,000 people in an area of nearly 3,500 square miles. Thick forest blanketed much of the land. "The country generally is only inhabited round the shores of the creeks & Bays," Darwin wrote, "the road by the coast is in some places so bad that many houses have scarcely any communication with others excepting by boats."

Darwin saw the island's inhabitants as impoverished. Though they lived surrounded by food and had adequate clothing and plenty of firewood, they could scarcely afford European luxuries like sugar, gunpowder, tobacco, or accurate timepieces. In the margin of his journal, he wrote with astonishment, "No Watch or Clock, strike the Bell by guess!" He also recorded his distaste in a line bound to give anyone reading his journal today whiplash: "Besides the Climate, it is disagreeable to see so much poverty & discontent. Poverty is a rare sight in S. America."

Poverty, the Dickensian kind Darwin would have known from London, is an all-too-common sight in modern South America, so much so that in places squalor has become a tourist draw. The slum tours, though, are mostly an urban phenomenon. In countryside areas like Chiloé, poverty had not yet acquired the glamour it had taken on in Rio de Janeiro.

Certainly, the farmers of Chiloé were still poor, and many of them lived off the land, or, more accurately, the sea. Van Meurs, the history museum director, told us that modernization had brought new and different jobs in the trinket shops and supermarkets of Ancud, but that it hadn't changed the basic culture. "People still maintain their ties with the land. They haven't lost their contact with the environment," she said. "Everyone goes fishing. The man who works at the market, he works a lot but everyone knows when he gets off he's going fishing."

Chilotans have lived off seafood since Darwin's time. Darwin specifically mentioned their *corrales*, underwater hedges used to trap fish at low tide, although Don Carlos told us that modernized fishing techniques have mostly eliminated the traps. As Josh and I passed along the shore at low tide we saw colorful yellow-and-red fishing boats resting on the sand, waiting for a high tide to come and float them. Chiloé didn't seem to have much of a commercial fleet, just a collection of small sixteen-foot wood-plank private boats, to be used whenever someone got hungry. Don Carlos had spent nearly ten minutes explaining to us how to make the island's specialty food, so after visiting the museum we went to sample *curanto*, a dish of mussels, clams, potato-and-flour cakes, and small slices of pork. We chose a restaurant overlooking the wharf, on a street painted by Martens in 1834. He had captured a dirt street and two-story wooden houses; now, convenience stores, restaurants, and a Goodyear tire shop, mostly built with concrete and corrugated metal, lined the paved route. I ordered the *curanto* and Josh ordered a salmon *ceviche*. "I think we're basically breaking the first two huge rules of eating abroad," Josh said as the waiter brought out my steaming plate of shellfish. "Mussels and raw fish—it could be a long night."

It wasn't. Soothed by the rhythm of a torrential downpour on our tin roof, we went to sleep early and woke up the next morning refreshed and ready for a hike.

On their first trip to the island, after a few weeks in Ancud, the Beagle's crew had seen enough of winter in Chiloé. On July 13, the ship weighed anchor and tried to beat its way out into the Pacific against a rough swell. "We were all glad to leave Chiloé; at the time of year nothing but an amphibious animal could tolerate the climate,"

Darwin wrote in his diary. But, characteristically optimistic, he looked forward to returning: "In summer, when we return, I dare say Chiloé will wear a more cheerful look."

He proved prophetic. When the Beagle returned in November, "the island wore quite a pleasing aspect, with the sun shining brightly on the patches of cleared ground & dusky green woods." After two months of survey work around the islands, Darwin set off on a trip to the rain-battered west coast of Chiloé, riding for Cucao, then and now the only inhabited part of the island's west coast. Josh and I trailed along on an early morning bus. It rattled along on a winding gravel road, bumping past farmhouses and fields. We passed a massive commercial salmon farming operation called "Salmon Net" and then along the shore of a large lake.

A long cement bridge connected Cucao with the road to the rest of the island. A few closed-and-shuttered restaurants, including one called the "Darwin Rest Stop" were scattered around the entrance to the Chiloé National Park, which covered much of the coastline between Cucao and Ancud to the north. Josh and I hopped off at the park headquarters and hiked out toward the ocean, crossing a stile over a barbed-wire fence.

As we walked, the thick, dark forest faded into a series of sloping gray dunes dotted with scrubby low bushes and reedy, tough clumps of grass. Heavy clouds took most of the color out of the vegetation, rendering everything in shades of gray. The weather seemed indecisive, unsure whether to rain or not, overcast skies misting and dripping but never pouring or drying up. I mentioned this to Josh and he disagreed. "It looks pretty decidedly overcast to me," he grumbled.

I didn't expect much of a welcome from the water. Darwin had written of the "terrible surf" which on stormy days could be heard at night from Castro, twenty miles away. Arriving where we now walked, he rode north a little and then abandoned the pursuit in the face of impenetrable forest and rugged coastline, "fronted by many breakers on which the sea is eternally roaring."

The dunes reached back about a quarter of a mile from the beach, and Josh and I stumbled across the hard black sand. Bits and pieces of trail wound through the sparse vegetation and dipped into swampy puddles in the low areas. The flat spots in the dunes stretched off in a shallow valley that had the style of post-apocalyptic nuclear wasteland paintings. Stunted, blackened plants curled over like burned stumps amid the wind-swept sand and a few large, drab-olive plants that looked like four-foot-wide maple leafs growing out of the ground. For Darwin, this would have felt like reaching the end of the world again, yet another place like Tierra del Fuego or the rainforest to reinforce how far he was from England. For us, still only four days removed from home, and an hour-long bus ride from pizza restaurants and Internet cafés, it felt a little different.

When you're traveling, sometimes you want so badly for your own trip to fit into everyone else's trip. Josh and I wanted the Chiloé National Park to feel foreign and lost and isolated, the way it had for Darwin. We wanted the

Magic Isle to show up and earn its title. Darwin wanted Patagonia to feel vast and romantic and exciting, the way it seemingly had for other travelers. Well, you can't always get what you want. Sometimes it takes a while to sort out your own mind, as it did for Darwin in Patagonia. Sometimes, modern life just hoofs right up and moos in your face, as it promptly did to Josh and me.

After a last small rise in the dunes, the beach sprawled out in front of us, a long sandy crescent stretching off into a cloud of spray flying off the backs of the breakers. Tugged by a serious rip current, the tidal zone frothed in a confused, seething jumble of foam. I poked around at the top of the dunes, analyzing the currents and the waves and the types of breaks. Josh interrupted my contemplation. "Say," he said. "Are those cows?"

I looked in his direction and saw him crouching, sneaking up on a brown and white heifer for a picture. The cow was standing, looking moodily at Josh and slowly masticating a bit of sandy grass. "What are cows doing on the beach?" I yelled back. "Are they actually eating this stuff?" I kicked the stiff, reedy grass, which brushed off my boots and refused to be trampled. Josh ignored me. He had spotted a solo cow trudging down the beach and chased after it to get that perfect, paradisiacal vacation shot of hoof prints in the sand on a lonely beach. The cow obliged by walking slowly, swaying a bit from side to side as if lovesick and distracted. I trailed along, noticing more cows concealed behind every dune. Sometimes they reared up out of nowhere, hidden perfectly by the shifting sands and camouflaged by the gray landscape. It was entirely, completely ridiculous.

So much for isolation and magic in this corner of the island, I thought. One myth of Chiloé deflated.

We tried again the next day. Josh was eager to visit a small town on the opposite shore of Chiloe called Quicaví. Supposedly this was the headquarters of a cabal of witches. How the witches fit into the rest of the island's mythology was never made entirely clear to us; they weren't exactly a part of the strange creatures stories—after Trauco and Fiura, witches just sound so . . . normal—but then, they weren't exactly separate. Some books blended the two, claiming that the witches used another Chilotan legend, a ghost pirate ship known as Caleuche, as their chariot. Don Carlos attempted to explain it to us by saying that people in different parts of the island celebrated different things, which ended up sounding like people just kind of told stories of whatever personal superstition they felt like telling. "Go to Quicaví and ask them about Trauco!" he said. "They don't know!"

But they did have the witches. Bruce Chatwin, the author of *In Patagonia*, reported staying in a hotel in Rio Gallegos that was full of workers from Chiloé, and they told him all about it. Chatwin said there was a group of witches called the Recta Provincia, dedicated to hurting ordinary people. They had two main offices, in Buenos Aires and Santiago, but smaller regional offices (I'm paraphrasing a bit here) scattered around the countryside, including the cave in Quicaví. It wasn't easy to visit, though. "Any visitor to it suffers thereafter from temporary amnesia," Chatwin wrote. "If he happens to be literate, he loses his hands and the ability to write."

(Obviously, we didn't make it all the way.)

In fact, we hitched a ride to Quicaví, walked up to the first house we found, knocked on the door, asked the matronly woman who answered about the cave, and were immediately stymied.

"You can't get there anymore," the woman said. "It's caved in."

"Is there anyone around town we can talk to?"

She directed us to find an unpainted two-story house and ask for the retired schoolteacher who lived there. We thanked her, walked across the street, and found the house as described. Another old woman answered the door, and when we told her about our quest, invited us into the living room. She sat us down, poured us two cups of orange Fanta, brought us cookies, and told us to wait while she fetched her husband.

Marcelo Marcias emerged a few moments later and sat down opposite us at a wooden table. As he introduced himself I realized that while I had thought Don Carlos and Antonio's accents were difficult, they had actually been models of enunciation and clarity compared to this more rural Chilotan accent. If languages are digital things, with distinct words, then the best way I can describe Marcelo Marcias' language is to say that he spoke in analog. I asked him about the witches, and although I missed much of his answer, I did understand one important line.

"The people who told those stories have died," he said.

In 1883, Marcias said, the mayor of Chiloé conducted

a good old-fashioned witch-hunt. Witches were identified by their neighbors, and the mayor "killed all the witches."

"And what about the cave? People have told us it doesn't exist."

"*Seguramenteenestamomentono*," he said, which I think translates to, "Nope."

Marcias said he thought he had an idea where the cave might have been, and when we looked later, we found newspaper articles that claimed to know of *a* cave, if not *the* cave. But even that cave was off-limits now, separated from Quicaví by a treacherous ravine and a washed-out road.

I asked Marcias if he thought the witches still existed.

"Oh, yes, they exist," he said.

"So what would happen," I asked, "if the witches met Trauco?"

"Oh, that's not very likely," he said. "The witches have far more power. Trauco is weak."

At night, he said, people would see lights out on the water—the lights of Caleuche ferrying witches around. Josh and I didn't plan to stay the night to look for lights on the water. "So the mystery of the witches remains," he said later. "I sort of like it better that way."

Before we left, we went to visit one last trail: The *Senda Darwin*, a small biological station on the north shore of the island. The morning dawned beautifully. A full rainbow spanned the green fields, and a second half-rainbow stretched up behind it. We could see right to the end of both. "There's no pot of gold," Josh later declared. "That's another capitalist lie."

We found a ranger, Emer Mencilla Díaz, standing out front of a small visitor's center. He invited us in, and the second we crossed the threshold it started pouring outside, drumming on the windows and roof and splashing up from puddles in the mud.

"Is it like this all the time with the weather?" I asked. "A little sun, a little rain?"

"Yes." Emer paused, smiled, and corrected himself. "Well, a lot of rain, a little sun."

We looked outside. The sun had come out, but the rain still fell. A small flock of sheep across from the visitor's center had turned black about halfway down their coats from the mud. At this point, they looked rather as if they had turned white from the necks up. Emer told us that the station used to be a study spot for biologists with an office in Santiago but that eventually they had decided it would be more convenient to just move in—so they bought the land and named it after Darwin.

He offered, when the rain cleared, to take us along the Darwin trail. Like Don Carlos, Emer was a lifelong resident who liked talking about the island's history. But instead of mythology, his preferred topic was island ecology.

"It's changed a lot, of course," Emer said. He told us he had been born in Quellón, on the very southern tip of the island, and seen firsthand the changes. In particular, he said, logging and burning had erased the once-forested face of Chiloé, transforming it into pastureland and open space, while salmon farms had polluted the water. "It destroys everything we have," he said. "But for some people a lot of money means the environment isn't important."

He looked outside. The clouds had moved on, leaving the sun shimmering on the water leaking out of the saturated ground. He invited us to come along and walk the Darwin trail.

"Tourism people are always fighting," he said, putting on a coat. "'Darwin passed through here,' 'No, through here.' One old man is absolutely sure he had Darwin's footprint. This part here, the elders say there was a trail here, so this is where Darwin went." We walked along a ridge, with the scientists' guesthouse on the right and a shallow, heavily forested canyon on the left.

Emer led us past a farmhouse where he told us biologists test different kinds of seeds to see how quickly they grow. We slopped along through the mud and clipped grass, Josh battling away thorn bushes with inch-long auburn thorns that stuck to his pants. We took a turn into the forest. "It's really difficult to walk through the forest when there's no trail," Emer said as we plunged into ferns and bamboo. "Very difficult."

Emer walked up front with his hands raised to push spider webs out of the way. He occasionally dove into the forest to retrieve some bit of moss or leaf or spider to bring back and show us. As I watched Emer leap into one of the shin-high puddles in search of a kind of streaming algae-like thing he identified as *pompón*, I felt a happy kinship with Darwin: Exploring a strange new world with a friendly local guide, wondering about the environment, and diving into puddles with reckless disregard for personal dryness in search of an interesting plant specimen.

Darwin was, like all his shipmates, happy to leave the wet and cold. "I believe every one is glad to say farewell to Chiloé," he wrote in his journal. But he didn't give up on a place that easily, and he finished off with a more upbeat view: "Yet if we forget the gloom & ceaseless rain of winter, Chiloé might pass for a charming island."

As Josh and I stood on the ferry that afternoon, looking back over the island shining in the sun, and the rainbows scattered across the sky, and the peaks of the distant, snow-covered mountains, it was hard not to feel a bit reluctant about leaving.

Finally. Score one for the Magic Isle.

12: VALDIVIA

The Apple Story

> *The uniformity of a forest soon becomes very wearisome; this West coast makes me remember with pleasure the free, unbounded plains of Patagonia; yet with the true spirit of contradiction, I cannot forget how sublime is the silence of the forest.*
>
> —BEAGLE DIARY, FEBRUARY 12, 1835

DARWIN SAILED NORTH IN FEBRUARY 1835 from Chiloé to Valdivia, a small town snuggled onto a bend at the confluence of two rivers, roughly ten miles inland from the coast. The Beagle anchored at sea under the ruins of some old Spanish fortresses, and Darwin rode in a smaller boat up the river to the town. He reported that the scenery, apart from a few Indian huts, "is one unbroken forest." When they arrived in the town itself, Darwin found it "completely hidden in a wood of Apple trees; the streets are merely paths in an orchard. I never saw this fruit in such abundance."

Josh and I took the bus north from Chiloé and arrived in Valdivia in a hammering rainstorm. We'd been lucky with the weather in Chiloé. We were not in Valdivia.

I read out loud from Darwin's diary as we passed through cleared farms and into the outskirts of a university town of about 120,000 people. "It seems like it's added a few buildings since Darwin's time," Josh said. Instead of unbroken forest and apple orchards we saw farmhouses, meadows, and then dusty warehouses and electronics superstores as we moved through the city. Active clues to the forest's disappearance swayed along the road in front us—a series of logging trucks crept through the rain, hauling long, telephone-pole-sized logs behind them.

We drove through town and found a hostel with a nice view of the river. The Argentine owner had studied journalism for a year at San Francisco State University and invited us over to his riverfront house for pisco sours. He told us that he'd love to show us around, but he was leaving for Tibet in a day and had to pack.

I questioned him briefly about Darwin's visit, without much luck, so I switched to Valdivian history. "Where did all the apple orchards go?" I asked him. Darwin, in addition to labeling Valdivia one big orchard, had penciled an intriguing side note in the margin of his diary: "apple story." I wondered whether he meant something specific that had happened to him? A local legend about apples? The Chilean version of Johnny Appleseed? Like the cows on the beach, though, we seemed destined to be stymied in our quest for explanations.

"What apple orchards?" Lionel the hostel owner asked.

"There used to be apple trees everywhere here," I said. "That's what Darwin wrote."

He shrugged. "Maybe you can ask at the museum?"

The next morning, I woke up early and went downstairs and found Josh already standing at the door. "I can't wait to go out," he said morosely as he stared out at the rain. We sat down for breakfast while the rain drummed on the roof. "We used to have days like this in Santa Barbara," Josh said, "and no one would leave their house. No one would go to class. We would just take a rain day." He pulled his tea mug closer to him and wrapped his hands around it. He looked up at me, determination written in his clutching embrace of the tea.

"I'm just going to have a leisurely breakfast," he continued.

"OK," I said. I was just as determined that we were going to go ask about apples at the museum. Eventually, I felt, this constantly getting referred to someone else would pay off. But there was no point in starting an argument. Yet.

After another hour, Josh ran out of tea, and defiance, and we set off for the museum, a converted mansion with downstairs rooms devoted to the more famous personalities in Valdivia's lengthy history. The most outsized of these personalities seemed to be Admiral Lord Thomas Cochrane, a Scottish lord whose gold-tipped cane rested national-archive-style on a plush red velvet pillow. Cochrane had led an assault on the Spanish stronghold at Valdivia as part of the Chilean war of independence.

"This port is well known from Lord Cochrane's gallant attack when in the service of La Patria," Darwin wrote upon a rriving. Cochrane and his troops pillaged Valdivia, and the new Chilean government punished the town for its Spanish loyalties by not rebuilding. By the time Darwin visited, it had shrunk in size and its economy was stagnant. A number of escaped English convicts from Australia had recently landed in Valdivia and were welcomed with open arms. (Darwin noted that they all were married within a week of arriving). "The fact of their being such notorious rogues appears to have weighed nothing in the Governors opinion," Darwin wrote, "in comparison with the advantage of having some good workmen." Lithographs from 1835 showed a church plaza and a few scattered houses overwhelmed by a thick band of forest. The once-impregnable forts seized by Cochrane were rotting into the ocean. Darwin visited one called Niebla, where the cannons appeared in such bad shape that they would probably disintegrate after one shot.

The museum was empty and quiet. The only other person was a museum guard who stood outside, under an awning, watching the rain. His desire for society did not extend to telling wild-eyed, dripping strangers about apples. He told us that maybe we could come back tomorrow. We asked him how to get to Niebla. Yes, of course, he said, buses, just down at the street and stand at the stoplight and certainly you'll find one, won't be long at all. Josh looked peeved. Something no doubt about the steady succession of people who had, for the better part of the last twenty-four hours, been keeping him from his tea and the hostel fireplace, and instead leading him ever onward so that he could stand around waiting in the downpour.

"Somehow," Josh said as we left, "I think that guy is behind us cackling like mad." We bent over and leaned into the wind and splashed toward the street. A fire alarm went up across the river, followed by cathedral bells ringing.

"I wonder if the cathedral is on fire," I said.

"I think anything on fire would be pretty quickly put out in this weather," Josh said.

"I was thinking we might warm up first."

Josh switched to his radio voice, a more melodious, rhythmic version of his normal wisecracking. "And about halfway through the trip," he joked, "Eric came out in favor of church burning."

We waded down the street to the corner, where a bus for Niebla promptly picked us up. The bus radio played Lionel Richie's "Dancing in the Streets." We drove along the river, gray and confused, with whitecaps like the ocean. The rain slammed the side of the bus so hard we couldn't see out the windows. At the end of the line, the driver indicated we should get off. "Where's the fort?" I asked.

"Oh, just up the road a bit," he said, without any hint of the malice such understatement implied. He jerked a thumb back where we had come from.

We set off walking. "I think he meant just a block or two," I said, trying to keep Josh's confidence up. After two blocks, I said I figured it would only be a short bit more. After a short bit more, we ducked inside a convenience store to ask directions and linger for a bit while drying off. I looked at Josh. He hadn't brought a hood with him—"I thought it would be too much of a hassle to pack," he explained—and water had collected in a pool on the collar of his jacket and was running down the back of his neck. His thin, nylon pants stuck to his legs, defining the contour of his knobby knees. Water rushed inward from there and drained down into his waterproof hiking boots. He had his head down and wore the same expression as a horse when it's waiting out a storm in an open pasture. Little drops of water dripped off his nose and beaded around his cheeks. He looked absolutely miserable.

It was my turn to cackle. "Just at the top of this hill," I promised.

We found the fort on a high bluff overlooking the river and ocean. It looked about the same as it would have in Darwin's time, except that one of the stone buildings had been converted into a museum. A guard, huddling away from the rain at the entrance, took our entrance fee and asked us not to walk on the ruins. We promptly crossed over to the old guard house, skipping most of the fascinating but wet outdoor foundations. Inside, we found maps and flags and a small replica frigate. Although it mentioned nothing specific about Darwin's visit, a text display suggested the fort had been mostly left alone from the time of Cochrane's assault until restoration efforts in the 1900s. An elderly eight-pound cannon rusted in place, still pointing out at the spot near the opposite riverbank where the Beagle had anchored. We watched through slits in the walls as the wind and rain sped by like jet blasts. A few feet of grassy bluff beyond the fort tumbled into the heaving gray ocean, which dissolved into the equally gray sky.

"I saw a painting once that looked like this," Josh said. "It was called 'Vertigo' and it had a naked woman standing at the end of a pier, and the ocean and sky blended together."

"Well," I pointed out, "they don't blend together perfectly because you can see the whitecaps on the ocean."

Josh looked around forlornly. "And I don't think a naked woman would survive very long out here," he said.

Josh appeared to need some cheering up, or at least dry socks and a boisterous night out. Back at the hostel that evening, we found a group of recent college grads from Vermont who were celebrating a birthday and talked them into dinner at the Kunstman Brewery—the proud location of the original brewery started by a German immigrant named Philip Andwandter in 1850, where they had been making the best beer in Chile uninterrupted for the last 150 years. The brewery's menu promised beer that was pure, great-tasting, and "nutritious," made in accordance with the strict purity laws of Wilhelm IV. We took a taxi.

We ordered two *columnas*, three-foot-tall lab-style graduated cylinders filled to the 2.5-liter mark with beer. Each had a half-foot-tall wood base and a tap, and measurement markings up the sides.

"Did Darwin visit the brewery?" the graduates wanted to know.

"It wasn't here when he was here," I explained. "Although he might have, if it had been." I shared my conflicting views, which I'd been pondering since Buenos Aires, on Darwin the prude versus Darwin the partier. By most accounts, Darwin was a bit of a boring guy when it came to the social scene. He didn't like to dance and wasn't particularly lively or glib. Much more Mr. Darcy than Mr. Wickham. On the other hand, he'd had a sort of unspoken engagement understanding with a girl back in England before he left. She used to write him rather suggestive letters and talk about making a "beast" of herself in the strawberry beds with him. Then he'd sailed away and reached Brazil a few months later where a letter from his sister was waiting for him that said, hey, you know that girl Fanny—guess what? She's getting married! So Darwin was upset for a while, had a good cry in the rainforest, and then he was single for the next five years. A twenty-three-year-old on a long voyage with frequent city visits in the company of a bunch of sailors. But then, I told the Vermonters, Darwin wasn't a sailor. He was a gentleman. He was along on the trip primarily because FitzRoy needed another aristocrat to talk to.

"I mean," I said, "he *was* English."

"I had always kind of pictured him dropping his monocle in his wine," said one of the grads. "'I say old chap,' that kind of stuff."

"He was never like that, I don't think," I said. "He walked down the streets of Buenos Aires making jokes about the 'signoritas.'"

In fact, Darwin did mention drinking in Valdivia. "An old man illustrated his motto that 'Necessidad es la Madre del invencion' by giving an account of how many things he manufactured from apples," he wrote. "After extracting the cyder from the refuse, he by some process procured a white & most excellently flavoured spirit (which many of the officers tasted); he also could make wine." No hints though as to whether he had dropped his monocle in it and said "I say old chap."

And once again, apples.

Josh showed little inclination the next morning to face the rain again, so I set off alone to find out what I could about Valdivia's disappeared apple orchards. The museum guard had told me to come back and ask, but had mysteriously neglected to mention that the museum was closed for the day. I puzzled for a while at the locked doors, thinking, *but he told us to return*! The locked doors appeared unsympathetic. I started back toward the hostel. As I stumbled across the river bridge in the rain, I looked down and saw the big-top tent of Valdivia's fluvial market. Sea lions barked and begged for fish at the edge of it, but in the middle, I was fairly certain, there were produce vendors. I wandered in and found a man selling apples, and bought two.

"I read a book that says all of Valdivia used to be covered in apple orchards," I said. "What happened?"

The vendor stared at me blankly. "What?"

"Where are all the apple trees?" I asked.

"Oh," he said. "They're just outside the city now." He waved off into the distance.

On February 20, 1835, Darwin decided to take a nap in one of those forests in the distance. He settled down and suddenly, the earth started shaking. According to his diary, he behaved in a manner becoming an English scientist, coolly rising to his feet to measure the direction of the earthquake. "There was no difficulty in standing upright; but the motion made me giddy," he wrote in his journal. "I can compare it to skating on very thin ice or to the motion of a ship in a little cross ripple."

Darwin didn't realize the extent of the quake until he returned to town. In the forest, he had been isolated and far from buildings, and as he recorded it, the breeze stirred the trees and the earth rumbled and that was that. "It was a highly interesting but by no means awe-exciting phenomenon," he wrote. In Valdivia, however, the houses had shaken violently until many of the nails had come out, and seeing the dread on residents' faces convinced Darwin that it was considerably more awful in town. But Valdivia had been lucky. Further north, near the epicenter, an entire city lay in ruins. In what would turn out to be a considerable understatement, Darwin wrote, "I am afraid we shall hear of damage done at Concepción."

13: CONCEPCIÓN

Shaken, Not Stirred

> *To my mind since leaving England we have scarcely beheld any one other sight so deeply interesting. The Earthquake & Volcano are parts of one of the greatest phenomena to which this world is subject.*
>
> —BEAGLE DIARY, MARCH 5, 1835

IT TOOK THE BEAGLE NEARLY TWO WEEKS to sail one hundred miles north to Concepción. When they arrived in the harbor on March 4, an estate owner rode down to the ship and told them that the earthquake had destroyed everything and that in Concepción and its port town, Talcahuano, not a house remained standing. Darwin went out riding and soon found proof. "The whole coast was strewed over with timber & furniture as if a thousand great ships had been wrecked," he wrote. "Besides chairs, tables, bookshelves &c &c in great numbers, there were several roofs of cottages almost entire, Store houses had been burst open, & in all parts great bags of cotton, Yerba, & other valuable merchandise were scattered about." On a small island in the bay at Talcahuano, Darwin recorded huge cracks in the ground and rocks covered in sea life cast high onto the beach. FitzRoy conducted his own studies by measuring mussels and seaweed on rocks that had once been underwater and concluded that the earthquake had lifted the rocks ten feet.

The next day the captain and the naturalist rode through Talcahuano and Concepción to survey the ruins. All of Concepción lay in heaps of bricks and timber, and debris clogged the streets, forcing Darwin to climb over piles several feet high to get from place to place. "The ground is traversed by rents, the solid rocks are shivered," he wrote in a letter home. "Solid buttresses 6-10 feet thick are broken into fragments like so much biscuit."

The earthquake had taken place at 11 A.M. and many residents had been outside, which probably saved thousands of lives. Darwin interviewed the English consul, who told him that he felt the first motion of the earthquake while eating breakfast and immediately ran outside, but only reached the courtyard before his house began to collapse behind him. The consul climbed atop part of the house that had already fallen, but the motion of the ground prevented him from standing. As he crawled up the pile of debris, the rest of the house collapsed. "The sky became dark from the dense cloud of dust; with his eyes blinded & mouth choked he at last reached the street," Darwin wrote. "Shock succeeded shock at the interval of a few minutes; no one dared approach the shattered ruins; no one knew whether his dearest friends or relations were perishing from the want of help. The thatched roofs fell over the fires, & flames burst forth in all parts; hundreds knew themselves ruined & few had the means of procuring food for the day. Can a more miserable & fearful scene be imagined?"

Darwin guessed that "not more than 100" had died, although many others still lay buried in the rubble. In his longest diary entry from South America, Darwin described the ruins, theorized about the directional origin of the earthquake, wondered what would happen if an earthquake struck England—"the earthquake alone is sufficient to destroy the prosperity of a country," he wrote—and delved briefly into earth science. He wrapped his musings up with a revealing comment about his own personality and the clash between his generally sympathetic nature and his lust for novelty—and geology. "It is a bitter & humiliating thing to see works which have cost men so much time & labour overthrown in one minute," he concluded. "Yet compassion for the inhabitants is almost instantly forgotten by the interest excited in finding that state of things produced at a moment of time which one is accustomed to attribute to the succession of ages."

Concepción, for the moment, is rebuilt. We expected to find sturdy, defensive-minded architecture, since major earthquakes had destroyed all or most of the city in 1570,

1657, 1835, and 1939. The largest recorded earthquake in human history, which struck offshore from Valdivia in 1960, caused by comparison only mild damage in Concepción. The subsequent tsunami, however—which also hit Japan, Hawaii, and the California coast—was devastating.

Josh and I arrived in the evening at a concrete bunker-style bus station in the middle of a rainstorm. We found a flea-ridden, dirty hovel on the edge of town and settled in for the night. Josh's sagging bed cracked as he sat down. Within minutes of lying down, small welts began to appear on my arm and back, bringing to mind something Darwin had written about camping near Valdivia. "Our resting house was so dirty I preferred sleeping outside," he wrote. "I am sure in the morning there was not the space of a shilling on my legs which had not its little red mark where the flea had feasted."

Sometimes, in romanticizing Darwin's trip as a wonderful adventure, I forgot that travel in the nineteenth century had certain blood-sucking disadvantages as well.

"I think something's biting me," I informed Josh.

He raised an eyebrow.

"Actually," I continued, "I *know* something is biting me."

We slept poorly. The hostel was full of laborers who woke up early to take breakfast and go to work. Someone started hammering something in the hallway outside our room at 7:15 A.M. I struggled out of bed and wandered down the hall. The bathroom was a miserable hole, with a rusting, dirty toilet and a shower that looked like something out of the better class of mystery novel. The shower had a huge crater in the middle of the tile, with a small metal grate not doing much to conceal it. There was a picture of a seahorse on the grate, though. I decided to skip the shower and trudged back to the room, where I flipped on the lights.

"I think they're building an airplane out there," Josh groaned.

"Let's go somewhere else," I said.

"I fully expect to walk out and see an assembled light craft in the hallway," Josh said.

We switched hotels and found, to our surprise, that the rest of Concepción was not at all a miserable place. The University of Concepción, one of Chile's finest, lent a good vibe to the city, small, high-character cafés stayed open late and proudly offered real coffee, and numerous clean, leafy plazas kept anything from feeling too industrial. The Plaza de Armas had a beautiful fountain in the middle and a nice, earthquake-safe cathedral on one end. The cathedral's towers fell down in the 1939 earthquake (and earlier, in 1835, etcetera), and the citizenry evidently decided that they had watched their towers crumble enough times. So for now, no towers.

The 1835 quake, we discovered, was known as "*La Ruina*." Concepción had suffered so many that each had been given a nickname.

We hiked over to ask the director of the regional history museum about earthquakes. It was starting to get a little bit old, going to museums in almost every town, but they were such useful little places, collections not just of local history but of local values, revealing so much about what the town prioritized. Stuffed birds and sheep shearing photos in Patagonia, a small display disparaging the native people's keenness in Tierra del Fuego, the gold-tipped cane of an old Scottish lord in Valdivia. The Concepción museum featured dioramas about Chilean history. There were several depicting Indian raids and battles, and finally near the end we found the earthquake diorama. Although it ostensibly showed the aftermath of the 1939 quake, it listed all the others, possibly on the assumption that a ruined town is a ruined town. Only the costumes and wrecked store signage would differentiate an 1835 pile of rubble from a 1939 pile of rubble. "The earthquake and tsunami known as *La Ruina*," a small plaque read, "partly changed the physical geography of the place where its ferocious whips were felt for three days. This new disaster took place in Concepción February, 20, 1835, and included the zone between Coquimbo and the islands of the archipelago of Chiloé."

Coquimbo and Chiloé are both 400 miles from Concepción, which means the scale of the quake was mind-boggling. Imagine an earthquake centered in San Francisco that's felt along the entire California coast, or a quake in New York City shaking Montreal, Canada, and Richmond, Virginia. Various experts estimate the magnitude of *La Ruina* at 8.5 on the Richter scale, which would make it easily one of the top ten largest quakes in human history. One of the reasons the Richter scale is no longer in official use—at least in the United States—is that it's fairly useless beyond a magnitude of about six, so who knows, really, what 8.5 means exactly. Since earthquake measurement didn't start until the early twentieth century, a more exact magnitude wasn't recorded. But that 8.5 estimate explains why Darwin was so fascinated, and why he struggled with feeling lucky to have been present. He had felt an earthquake so large as to be almost unimaginable. Most European geologists would have given their careers to witness such an event. And yet he also saw men crawling over ruins and entire families made homeless and hungry.

As we pondered the Playmobil-style men in the diorama, a museum guard approached and indicated that we could see the director now. He led us upstairs to a small corner office where a very large man and a smaller, Smithers-like man sat facing a computer. It was easy to tell who was who.

Without looking up or facing us, the director asked us to wait a few minutes, while he and Smithers scanned photos of Concepción from the 1940s. Proportionally, the museum director put Santa Claus in mind, though he sported a nicely trimmed gray beard and the standard professor costume of vest and bowtie. A golden breastplate and sword hung on the wall, and I asked his back if they belonged to him. Without turning, he said yes, "They're only replicas."

"For wearing to parties?" I inquired. Smithers smiled, but the director didn't turn.

Finally, the last photo appeared on his computer screen. He swiveled in his desk chair and looked us up and down and asked where we were from. "I'm from California," I said. "And he's from New York."

"Yes," he agreed. "I can see that by how white he is."

He paused and chuckled. "And you've got dark skin. You're from the California sun."

"Mediterranean," I explained. "My mom's family is Greek."

"Ah," he said. "Excellent. I'm European myself. Alejandro Mihovilovich Gratz."

He whipped out a business card and I wrote down the name.

"So," I said, by way of introduction. "What can you tell us about earthquakes?"

"What we have here, in Concepción, more than anything else, is earthquakes," he replied. "This is a city of earthquakes."

"The town has been destroyed seven times in the last 500 years," I said. "Is there anything you can do to prepare?"

"I don't worry," he said. "You're never prepared for an earthquake. If there's another nine or ten, there's nothing you can do."

I tried again. Surely the town must have done something for seismic safety.

"How have people changed with so many earthquakes?"

He pulled out his coffee mug and set it on the edge of the desk. "They don't put their things here," he said. "They put them here." And he moved the mug to the middle of the desk.

Gratz settled back into his chair and started to tell us about the tidal wave that followed the 1960 earthquake, which had been so powerful it lifted livestock out of their fields and tossed them high up in the air. "And there was a cow," he said. "Moooo, you know?"

He glanced over at Josh, to make sure he understood. Josh nodded.

"And it was launched in the air, like this."

He paused again, held his hands up and traced an arc in the air following the cow's trajectory. I was reminded of *Monty Python and the Holy Grail*. Josh looked suitably impressed.

"It's kind of refreshing to see a bit of fatalism," he said as we left the museum.

"It's a jolly fatalism," I agreed. "It's like a darkly comic version of Santa Claus."

"I think in the United States we've lost a bit of that," Josh said. "We're so afraid of everything, and we spend so much thinking we can prevent and prepare for everything, and sometimes, there's just nothing you can do."

In the nineteenth century, the English were starting to feel that they could solve almost anything through technology. This attitude is obviously prevalent in the United States of the early twenty-first century. The earthquake reminded Darwin—as the museum director reminded us—that sometimes people really are helpless in the face of disaster. Although I enjoyed the museum director's what-me-worry attitude, and his laughing at fate, Darwin reacted much more somberly. He'd "never again laugh" at the power of catastrophe, not after he had thought about it and realized how fragile England's own infrastructure was.

One of Darwin's most prized possessions on board the Beagle was a copy of geologist Charles Lyell's *Principles of*

Geology. The book set out a case for why the earth had been shaped over millennia by the same forces familiar to ninteenth century scientists—erosion, wind, waves, volcanos, and earthquakes. Lyell and Darwin would become close friends after the voyage, bonding at first over the way Darwin set out with such intense interest to prove Lyell's theories. Lyell in some ways returned the favor, as it was he, more than anyone else, who encouraged Darwin to publish *The Origin of Species*. But in 1835, still nothing more than Lyell's number one fan, Darwin was laying the seeds of their friendship in his west coast explorations. After the wonders of rainforests and ostriches on the east coast, Darwin had gone to Chile and embraced geology. It was, he wrote at one point, practically all he could think about. Good thing, because as he headed into the Chilean north, rocks, and the stuff in them, became the biggest story in town.

14: PISCO ELQUI

Demon Cactuses

> *The foreground is singular from the number of parallel & extensive terraces; & the included strip of green valley abounding with its willow bushes is contrasted on each hand by the naked hills.*
>
> —BEAGLE DIARY, JUNE 7, 1835

A FEW WEEKS LATER, JOSH AND I drove east through the Chilean desert in a rented Fiat. We stuck to a narrow cultivated valley along a shallow, gravel-lined river, following Darwin's trail of geological exploration from Coquimbo. The landscape looked just like the kind of place you'd find a wide-eyed mineralogist racing around, yipping with delight, trying to grab one of everything. It also looked like the kind of place you'd find some fool mineralogist's bleached bones. Amongst all the rockiness, a few hardy cactuses stuck it out and added a splash of olive to a palette of earthy rusts, yellows, and grays.

Josh, sitting in the passenger seat, hadn't said anything in about twenty minutes. I glanced away from the road to see what he was doing and saw him fumbling furiously with some radio cables from his backpack. He muttered about short splices and receivers and bent over his cables again with a roll of electrical tape. He twisted a few things together, gently laid the tape over them, and suddenly Bob Dylan and Johnny Cash were singing a duet through our car radio—he had spliced three cables together and turned his iPod into a small broadcasting device.

Then he looked out the window.

"Demon cactuses!" he yelped immediately. "Oh my God!"

A grove of long-armed cactuses leaned along the road, bristling with lengthy, scary spines. "Unbeknownst to most horticulturalists," Josh trembled, "there are actually three kinds of saguaro. Those with short spines, those with long spines, and the holy shit! variety."

Cactuses are not my kind of plant. I fell into a cactus once, face first, on a family trip to Arizona. Enough for me; I'll stick to ferns and wildflowers. But Josh had more to say about the spines and rather extended himself on the subject. They reminded him of convergent evolution, where different species in different-but-similar places respond by evolving the same adaptations. They reminded him of cactuses in the American southwest. Could they be related? They looked related. He wondered how cactuses would have made it here. Or did they start out here? How would a cactus walk from South America to North America? Were these really related to saguaro cactuses? Had they been introduced? Was there a Johnny Cactusseed who traveled the world, distributing columnar plants with, as Josh called it, "spines that would make an iron maiden look like a deck chair"? It was an inspired, entertaining monologue. Josh is now on his way to becoming a professor, and to all his prospective students, let me just say that you are lucky. Ask him about cactuses.

Meanwhile we piddled along past tranquil little villages so cute they ought to have starred in Disney movies. The sun beamed down through clear desert air and pure blue sky. A gentle breeze kept the temperature warm, not hot. The glare reflected on glowing crimson-and-gold vineyards along the river and lit up the pink rocks. Tiny adobe houses with driftwood posts grew out of the hillsides, their boundaries marked by explosions of fragrant purple flowers and blooming jasmine. "We passed through several small villages; the valley was beautifully cultivated & the whole scenery very grand," Darwin wrote.

Josh and I wanted to find the spot where Darwin turned around and recorded that observation, in the Coquimbo valley "where the R. Claro joins the Elque." We understood this would be the meeting of the Claro and Elqui Rivers, but depending on the map, that meeting place changed. On some maps the Elqui stopped halfway down a narrow branching valley, on others the Rio Claro was formed by the junction of the Elqui and the Rio Turbio flowing out of the Andes, and on still other maps the Rio

Claro itself trickled clear back to the cordillera. We asked three different strangers on the street and got three different answers, and decided to just drive along and see for ourselves. Maybe, like Darwin, we could inspect the haciendas and mines along the way.

We passed river junction candidate one—the odds-on favorite—where the Coquimbo valley split off into the Elqui valley and turned south to follow a narrow road up the Elqui valley. Two more possibilities followed, but neither seemed impressive enough to be the meeting of two rivers. By this point, the valley was only a quarter-mile wide and the streams flowing through it were small and lined with rushes. The canyon walls closed in on the road, which wound side to side, seeking out level spots wherever possible. A few houses clung to the slopes. "It's cool how all the things here look baked," Josh said. "Houses, land, trees . . ."

"People," I interjected, as we passed a shirtless, tanned man hugging a shady spot along the road.

After half an hour we arrived in Pisco Elqui at the far tip of the valley, almost certainly beyond where Darwin traveled. We stumbled out of the car and went to inspect the town. It was adorable. The odd-sounding Pisco Elqui was not its original name, we discovered. Originally, the town was called La Unión. Then politics got in the way, and not just any kind of politics: Alcohol politics. We knew that Chile and its neighbor to the north, Peru, both claimed to have invented a kind of grape brandy known as pisco. Peru, Chile noted enviously, even had a town named Pisco.

In 1939, the Chilean government decided that this situation would not do. The legislature passed a law, and La Unión, where vintners grew Chile's finest pisco grapes, assumed the mantle of Pisco Elqui.

From the shade of a bench-lined plaza and whimsical church, we could see RRR—the distillery for Chile's oldest pisco. I remembered one of Darwin's more agricultural observations: "The figs & grapes of Elqui are famous for their superiority & cultivated to great extent." Darwin didn't linger; agriculture didn't much interest him and geology and mineral mines did.

We lingered. "We went looking for mines to inspect and couldn't find any, so we inspected the pisco factory instead," I said to Josh. "Come on."

We crossed into a big adobe building and found the tasting room.

"RRR is the oldest pisco in Chile," explained the man behind the counter. "Here."

He set two wine glasses filled with a translucent yellowish liquid down in front of us.

I took a cautious sip and felt the cilia rising in alarm in my throat.

"Whaaa," Josh sputtered, "that burns!" ("Liver solvent," he later labeled it.)

The man explained that when Spanish settlers arrived in Pisco Elqui, they started growing grapes for wine. But because of Pisco Elqui's wonderful climate of 320 sunny days a year, the grapes had a very high sugar content. (More sun means grapes produce more sugar.) When they made wine from the grapes, then further distilled the wine into almost pure alcohol, it retained some of its flavor. *Some* of its flavor. To the uninitiated, pisco tastes like Drano.

He took us on a tour of the pisco distillery, through the grape presses and stills and down to the wine cellar, where pisco aged in California oak barrels. In the cool darkness, he turned on a projector and showed us a movie, with English subtitles and set to inspirational music, about the history of RRR brand pisco. "The indigenous people had developed a culture perfectly suited to receive the Spanish settlers," one subtitle read.

Josh and I snorted in chorus, disturbing a few Spanish tourists. "The primitive Indians were perfectly suited to receive the settlers," Josh repeated under his breath. "It's like an anthropologist's nightmare."

I liked the next phrase, too: "RRR, the pisco that conquered everybody else."

The lights in the cellar came back on, and we wandered out into the bright sunlight. While Josh continued mumbling about primitive Indians, I caught up to the guide. "Did you know that when Darwin visited here, he wrote about the grapes?" I asked.

"He did?"

"Yes."

"Interesting," the guide said. "So did the French naturalist, Claudio Gay. He traveled here just after Darwin and wrote that these were the best grapes in Chile. It's a very famous valley."

"I'm not sure he made it all the way here, though," I said. "Darwin wrote that he turned around at the meeting of the Rio Claro and Rio Elqui. Do you know where that is?"

"Sure," he said. "The Rio Elqui flows out of this valley, and meets the Rio Turbio. Together, they form the Rio Claro."

We had heard this answer earlier, but suddenly it made perfect sense. To Darwin, following the Rio Claro up the valley from the coast, the turnaround point would be the junction of the Rio Claro and Rio Elqui—and he probably never knew that the Rio Claro didn't continue beyond that spot.

Josh and I left the distillery and got back in the car and drove north into the sinking sun, with the valley half shadow and half blinding light. The grape vines appeared to be on fire, speckled with flaming reds, oranges, and yellows and glowing in the saturated light. We reached the meeting of the Rio Turbio and the Rio Elqui in a small town called Rivadavia. Amidst small convenience stores and farmhouses, a handwritten sign tacked to one adobe building offered yoga and massage. "The primitive natives were perfectly suited to receive the American tourists," Josh cracked.

Pulling off the highway, we hopped out and stumbled across a thick gravel bed toward the river. The green, turbid Rio Turbio flowed in from the east, splitting around a gravel bar but not mixing with the dark, clear water of the Rio Elqui. For a few hundred meters, the two continued to flow side by side, white water and dark water, a river of chocolate syrup running down through a pool of milk.

"It's beautiful," Josh said. "I wonder about the origins of these, that they'd have such different sediment levels."

"I wonder that Darwin didn't mention it," I said.

"Unless it wasn't this way in his time," Josh replied. "It

would have to be something at the source, unless it's so polluted now that it's changed color."

We picked up a few rocks and skimmed them across the surface. Our tiny splashes carried up in the air and hung like whale spouts while the current raced on by. The sun sank lower and disappeared behind the mountains. With their heavy bases, glowing sandy coloring, and sharp peaks, the mountains looked a bit like Egyptian pyramids, if only someone had seen fit to come along and plant grapes in the desert of Giza.

Once again, I found myself walking in Darwin's footsteps, watching another sunset from a place largely unchanged since his time (except for the yoga). As Josh and I laughed and compared skipping rocks and delighted in the last remnants of the day's sun, I realized that finding Darwin sites had taken on another level of meaning for me. There's a danger in labeling someone a genius; it makes them inaccessible. Darwin the Genius is beyond the reach of sympathy. But Darwin the person—the one who stood and watched the sunset over this same river, the one who would happily join in with Josh and I in skipping rocks—well, he was a lot like us. He *was* us. His career-crowning idea of evolution by natural selection is a triumph of human achievement that sprang from the perfectly achievable endeavors of careful observation, meticulous note-taking, and joyous, boundless curiosity.

15: ANDACOLLO

The Gold Mine

> *I was glad to take the opportunity of weighing one of the loads, which I picked out by chance. When standing straight over it, I could just lift it from the ground, the weight was 197 pounds. The Apire had carried this up 80 perpendicular yards, by a very steep road, & by climbing up a zigzag nearly vertical notched pole.*
>
> —BEAGLE DIARY, MAY 12, 1835

BY THE TIME DARWIN REACHED the Rio Claro in northern Chile, he had been away from home for more than three years. He had been living rough for those three years, sleeping half the time on a cramped ship, the other half on hard ground. Although his pronouncements about enjoying the freewheeling gaucho lifestyle and sleeping in random pastures continued, he was growing weary, and wrote in a letter home, "I am tired of this eternal rambling, without any rest." He had dealt with traveling's evil trifecta of food poisoning, insect bites, and theft. As much as he was tired of sleeping on the ground, it was impossible to sleep in houses because of the fleas, those "ravenous little wretches" who left his "whole shirt punctured with little spots of blood" and his skin "quite freckled with their bites." He had logged thousands of miles on horseback and hundreds more on foot.

But Darwin's curiosity persisted. Many Chileans seemed surprised at his geological explorations. "This was sometimes troublesome," Darwin wrote. "I found the most ready way of explaining my employment, was to ask them how it was that they themselves were not curious concerning earthquakes and volcanos? Why some springs were hot and others cold? Why there were mountains in Chile, and not a hill in La Plata?" More than any single quote from his journal, more than any scientific discovery from the five-year voyage, this one paragraph is the reason Charles Darwin discovered what he did and is celebrated today. He loved studying the world around him and wanted to explain what he saw. He had the courage to ask.

Even Chileans who believed that geological study was useful had trouble believing Darwin's motivations for sniffing about. He must be searching for mines, many concluded. That's what nearly all of the other gringos in northern Chile were doing. A man out wandering the desert, looking at rocks—what could he be doing other than looking for silver and gold?

Darwin first noticed Chile's craze for mining while climbing a mine-riddled mountain in central Chile. He also toured a copper mine at the base of the Andes in Jajuel. The mines were purely extractive, and the melting was done elsewhere. "Hence the mines have a singularly quiet aspect to those in England," Darwin wrote. "Here there is no smoke or furnaces or great steam-engines to disturb the quiet of the surrounding mountains." While Darwin spent several days "scrambling in all parts" of those mountains, the miners who lived there didn't get to take advantage of the scenery. They worked dawn to dusk. Their food consisted of sixteen figs and two loaves of bread for breakfast, boiled beans for dinner, and roasted wheat grains for supper. "They scarcely ever taste meat," Darwin wrote with apparent sympathy, "as with twelve pound per annum they have to clothe themselves & support their families."

The story was much the same elsewhere. At a gold mine south of Santiago, owned by an American named Nixon ("to whose kindness," Darwin wrote, "I was much indebted"), Darwin was "struck by the pale appearance of many of the men." These men, he learned, worked deep in a shaft mine, carrying 200-pound loads of rock up a rickety wooden ladder. They were only allowed to see their families for two days once every three weeks and were fed only beans and bread. "They would prefer living entirely upon the latter," Darwin wrote, "but with this they cannot work so hard, so that their masters, treating them like horses, make them eat the beans."

Still the miners didn't complain, Darwin noted. There

was more money in mining than in farming, and with mining, there was the possibility of finding a new vein and striking it rich. Darwin told one story about a man who'd picked up a rock to throw at his donkey—and found the rock heavy with silver. He had discovered the most profitable mine in northern Chile. Men were so enthusiastic for mining that on Sundays they would head out into the desert with crowbars, looking for exposed veins.

Much of northern Chile still depends economically on mining. After deciding to keep the rental car for another day, Josh and I found a representative sample of the craze in the high hills outside Coquimbo, in a small, red-dust-covered town called Andacollo. The city limit sign welcomed us to town—population 4,000—and pointed to an old mine where tourists could go to see demonstrations of old gold-mining techniques. I pulled off the highway into the dirt, and an old woman walking by the small wooden visitor's center waved at me. "There aren't any miners here today," she said.

"Are there any real mines around?" I asked.

"Oh, yes, just up the road."

We drove to the mine office, a portable building resting on stilts in the red dirt. Once we'd passed inspection at the gate and guard station, we were ushered inside. No window views reminded us of the desert, and the humming air conditioning, fluorescent lighting, gray carpet, and picture-lined cubicles could have been ripped from the twenty-second floor of a Manhattan office tower. A secretary told us to sit down, and a middle-aged man in a pink collared dress shirt came out to greet us a few minutes later. He introduced himself as Roberto Pardo, the mine's finance director, and in fluent English asked what I was doing at the mine.

"I'm writing about Darwin—"

"I'm the missing link!" he said.

He beckoned us into his office, where he sat down behind a desk and pictures of his two sons in their ice hockey uniforms. "I lived twenty-five years in Canada," he said. "The company used to have its head office in Vancouver. But after Pinochet's dictatorship left Chile, I returned. I like it here now."

I asked Pardo how things had changed since Darwin had described the conditions at Nixon's mine.

"Quite a bit," he said, smiling. "When we built this plant in 1995 it was technically one of the most advanced in the world. Our mine exploitation is done with a fleet of huge equipment. This last month, we moved 35,000 metric tons of rock per day."

Josh and I sat back, awed. Not quite the same as miners hauling rock out on their backs.

"It should have been more," he said, almost to himself. "But we had two days of rain. Still, that's 15,000 tons of rock with mineral and 20,000 we call waste. We have to crush 15,000 tons a day, otherwise it's not economical. It sounds exotic, but gold is very expensive to mine. You make more money in an iron mine."

In Darwin's time, the miners carried their gold-bearing rock to a mill, where it was ground into powder. Although the equipment was vastly different, the process remained the same—at Andacollo, the rocks were taken from the pit to a series of gigantic industrial crushing machines. In the 1800s, miners washed the crushed rock to leach out the gold.

"How is that done now?" I asked.

"Cyanide," Pardo told me.

Cyanide use turns out to be industry standard for gold mining. You spray the rocks with cyanide solution, and the cyanide picks up the gold, grabbing it out of the rock, and then drains down into a big pool. Then you run some kind of reverse filter type thing on it, using something like carbon that will pick the gold out of the cyanide, and then you refine it a bit more and end up with solid gold. Future students of Josh: ask him about this.

"We use 180 tons of cyanide a month, highly diluted," Pardo said. "It's a one-percent cyanide-in-water solution. People hear cyanide and go 'oooh,' but really, if you drink this, it won't kill you. Seriously. We had someone drink it a while ago by accident, and he was fine. It might kill a small bird, maybe. But not even a seagull."

Pardo's phone rang. "I've got a conference call with the American investors who own the mine," he said. "But I'll tell someone to give you a tour."

A few minutes later, the mine's Risk Prevention Supervisor, Oscar Urbina Valdes, came out and rescued us from the office-cubicle hell. (I kept thinking back to Jean, the Brazilian forest guide, telling me, "I could not work in some place with four walls.") I took "risk prevention supervisor" to mean security guard, and that's what Oscar looked like. Stocky, muscle-bound, neck-challenged, and sporting a shaved head and goatee, hardhat and sunglasses, the guy looked seriously intimidating. But Oscar was polite and pleasant. He found us hardhats and polarized safety glasses, and we climbed into his air conditioned, four-wheel-drive truck.

"Some people look as though they belong in a hardhat," Josh said, after checking himself out in the car's window. "I am not one of those people." He climbed uncomplainingly into the back seat, allowing me to ride shotgun.

We never exceeded fifteen miles an hour, on a nice, flat road. "I'm driving slowly as one of our agreements with Andacollo," Oscar explained. "To keep the dust down. We try very hard not to have an environmental impact. Of course we have one. It is a mine. But we have lots of agreements with the town to reduce our impact."

Nearly seventy percent of the mine's three hundred workers came from Andacollo, he told me, so I imagined that the mine would be popular in town regardless of its environmental condition. But Oscar seemed to be trying to head off any arguments we might make about the mine's environmental, social, or health impact. Having never heard of the mine before we had arrived a few hours ago, I hadn't come prepared to argue.

I vaguely understood that cyanide-leaching mines were not very popular around the world because in some places the waste cyanide—the stuff left over after the gold has been taken out—leaks into the ground and, despite what Pardo told us, at this point it gets quite dangerous. I knew that large-scale business operations run by American investors in relatively poor areas of South America were often rapacious, ruthless, and deceptive, and that they chose these places because they didn't have the environmental regulations and wage laws that North American governments might impose. And I knew that corporate protestations of environmental right-doing were usually half-truths at best. But although Darwin's was a curious spirit, it was not an activist one, and I often felt the same. I wanted to know how the mine worked and I wanted to see it for myself. But I didn't feel like making a scene and was content to be impressed with high-tech machinery.

Josh and Oscar and I passed a quiet afternoon touring the mine. Oscar took us to the "pit," a half-mile-wide terraced gash in the earth layered with red and white-streaked rocks. At the bottom, two massive dump trucks with fifteen-foot-tall tires waited for full loads of stone. In a few hours, Oscar said, they'd take several more trucks down, and each truck would haul one thousand tons of rock up to the masher. Swinging his arm around, he showed us the pulverizing machines, where men in reflective jumpsuits and respirators—looking like the kind of action-movie extras who die in scores at the point of James Bond's stolen machine gun—monitored a series of conveyor belts. (Oscar, following this metaphor, would be the muscle-bound villain who gives Bond several tense moments over the crusher.) The rock entered three separate crushers until it had been mashed into pebbles three-eighths of an inch wide, then traveled along on another series of conveyor belts that dropped all of the pebbles into "the pile," a mile-wide mound of crushed rock. Sprinklers sprayed the pile with cyanide solution, keeping the rock black and glistening in the sun. A gentle mist blew off the sprinklers and toward our viewpoint, but Oscar seemed unconcerned. He even took us up to a sector at the top of the pile where the sprinklers weren't running for a better viewpoint. "Truck ascending the pile," he called into his radio. "Repeat, truck ascending the pile." It sounded very cool. A few minutes later we stood, surrounded by glittering black rock, eyeing the sprinkler hoses and looking down across the entire mine.

Oscar pointed out the second pit. "They'll start laying the explosives after lunch," he said. "Will you be here for four hours more? You can wait in the office."

Josh and I exchanged a look, but this attempt at nonverbal communication failed because we were both still wearing our reflective safety sunglasses. But neither of us wanted to sit at someone's cubicle for four more hours. We'd already had a fascinating day, and I now understood Darwin's interest in touring the mines of Chile: There's pleasure in seeing how things worked. By the day's end, Josh was explaining its geological aspects to me in minute detail. (Excerpt from *his* journal from that day: "At a modern open-pit mine, cores are drilled from the earth every four meters in a grid pattern and sent off to geologists for analysis. The hole each core sample is pulled from is assigned a GPS address, and scientists come back with descriptions of the rock type and density at all different depths for each of the holes . . .")

Even in the 1800s, mining incorporated a tremendous amount of knowledge. Human capital lay behind every advance, from the discovery of veins to the leaching of minerals. "The washing when described sounds a very simple process," Darwin wrote after seeing Nixon's gold mine, "but it is at the same time beautiful to see how the exact adaptation of the current of water to the specific gravity of the gold so easily separates the powdered matrix from the metal." Not his punchiest quote ever, but it precisely conveys the fascination. It's something akin to awe-inspiring, to get outside your own field and see other humans thinking cleverly and designing such complicated, useful machines. Walking back through the office, past desk-bound engineers analyzing rock data, I felt the same way: deeply impressed at the centuries of accumulated wisdom that allowed geologists with GPS and computers to pinpoint specific metals in the ground, remove specific chunks of rock with precise explosives, and then use a chemical cocktail to turn the rock into gold bars—all in a facility straight from the Bond set. Regardless of the environmental and economic politics of the mine's investors, its nuts-and-bolts operation was pretty amazing.

16: BELL MOUNTAIN

From the Pacific to the Andes

> *The setting of the sun was glorious, the valleys being black whilst the snowy peaks of the Andes yet retained a ruby tint. When it was dark, we made a fire beneath a little arbor of bamboos, fired our charqui (or dried strips of beef), took our matté & were quite comfortable.*
>
> —BEAGLE DIARY, AUGUST 16, 1834

DARWIN'S GREATEST OVERLAND ADVENTURE took him out of the coastal Chilean city of Valparaiso on a late-fall ride through the Andes into Argentina. His enthusiasm for the trip showed in a last-minute letter he dashed off to his sister Caroline in which he laid out his plans for beating the snow in the high mountains (leaving that very morning at 4 A.M.), offered his backup plan (begging for horses to take him to Potosí in Bolivia) and concluded, "I cannot write more, for horse cloth stirrups pistols & spurs are lying on all sides of me."

At the time, there were two passes between Chile and Mendoza, Argentina. The pass called Uspallata was more commonly used, while the Portillo pass, further south and nearer to Santiago, was "more lofty and dangerous." Darwin, his guide Mariano Gonzales, and an accompanying mule train set out for the Portillo pass. The route took them first through the Maipo River valley, then after two days up a steep ascent of the Andes, which formed a double barrier between the two countries. Darwin climbed the western ridge most of the day, following switchbacks up through bands of snow until he reached the crest. The view from the top repaid the trouble. In a crossed-out sentence in his journal, he wrote that this view, above all others, stood out in his memory, and he penned one of his most beautiful descriptions: "The atmosphere so resplendently clear, the sky an intense blue, the profound valleys, the wild broken forms, the heaps of ruins piled up during the lapse of ages, the bright colored rocks, contrasted with the quiet mountains of snow, together produced a scene I never could have imagined." Darwin, apparently, had wandered away from his guides for awhile—or possibly they just weren't interested in the view—and the solitude amplified his emotion. "I felt glad I was by myself," he wrote. "It was like watching a thunderstorm, or hearing in the full Orchestra a Chorus of the Messiah."

The rest of the trip went off smoothly. He made it to Mendoza and then crossed back to Chile a few weeks later via the safer pass. In an April 1835 letter to his sister he wrote, "Since leaving I have never made so successful a journey." He proceeded to summarize his geological findings, realizing that she would understand little and care less, but his enthusiasm carried him through a long paragraph before he recognized, "I am afraid you will tell me, I am prosy with my geological descriptions & theories."

Beyond the successful geologizing, the scenery stuck with him. In a letter to his college professor J.S. Henslow, who unlike Darwin's sister did have a strong interest in geology, Darwin emphasized how much he enjoyed the view. "It is worth coming from England once to feel such intense delight."

Portillo is now a ski resort, the most famous ski resort in Chile, where the United States Olympic ski team goes to practice in the North American summer. But I wanted to climb one more mountain. Darwin had trekked through the cordillera near the end of his overland travels, a full three years into his Beagle journey. Rather than drag, the time flew by. The views seemed fresh and new even though he'd now gazed across South America from peaks in every conceivable condition. Darwin kept climbing regardless of what other things he had to do because to summit a mountain was to renew the joy of his journey.

Optimism may be one of the biggest benefits of travel. When you spend all your time in a small area, trekking back and forth to work, getting all your news on the Internet, it's easy to think the world is a lot worse off than it is. Then you get out in it, even for a short bit, and you get a summit view or find a friendly person who cares about nature just like you do, and then even when you go home you remember: Hey, it's not all bad. We're really doing OK. I wanted that view one more time for myself before I left, and I wanted that view for Josh for the first time. Josh had been burning out in an office for the last few years, and he needed that mountain climb whether he knew it or not.

We picked a more accessible mountain. Cerro La Campana, or Bell Mountain, didn't achieve the lofty heights of its Andean neighbors to the east, but it was the highest mountain in Chile's coast range, at roughly 6,000 feet above sea level. Darwin had climbed it and enjoyed the way that, from the top, all of Chile had appeared as if on a map. Situated a few hours east of the major port town of Valparaiso, it was also possible to climb the mountain, hike back down, and be back in the city for dinner.

It was a popular trail, but climbing it as we did on the shortest day of the year meant that there were no other hikers around. Which was not to say they hadn't been there. Hiking through clumps of frail, red-leaved trees, we passed boulders covered in graffiti. The scrawled names, dates, and hometowns—"Houston, Texas, 2003" or "Patty y Miguel, 1996" or "Elvira, 1972"—got more dense as we climbed, until almost every rock had someone's name and yearbook quote on it.

"Who brings a can of spray paint hiking with them?" I asked—other than, obviously, the group of six from Valparaiso who had left their names in yellow paint on the rock next to me.

"I fully expect the summit to be solid graffiti," Josh said. He started to point out some more distinctive markers—like "The Cure" (1999) and "Metallica" (date unknown). "At least we know the benefit now of always carrying a can of spray paint."

I grabbed a rock and pulled myself over it. "You don't see 'Charles Darwin' written anywhere here, do you?"

Actually, we did. At the base of the peak, surrounded by more happy graffiti, we found a bronze plaque honoring Darwin's ascent and placed by the "Scientific Society of Valparaiso, the British Colony and his admirers." I wondered how Darwin would have rated this memorial, out of all the hundreds of bronze things placed around the world in his honor. He certainly would have liked that it was a plaque in his name that didn't remember him as an old tormented evolutionist and instead honored him for climbing and adventuring. He would have liked that it remembered him with one of his own quotes, translated from English into Spanish: "We spent the day at the top of the mountain, and the time has never passed faster."

Time did pass quickly for him, and it wasn't because his work was so imperative. He'd spent a few days on and around the mountain already and done plenty of geologizing. Once, on a different mountain, annoyed because he hadn't reached the top, Darwin wrote that, "all purposes of geology had been answered," so it wasn't really necessary to reach the summit. (Looming darkness and thigh cramps had nothing to do with it, you see; he had merely concluded his geological studies.) At the top of Cerro La Campana, Darwin didn't need to spend the entire time geologizing.

He was instead taking a moment to restore his pleasure in an arduous voyage, to remember why it was that he'd been so eager to leave England, and to remind himself that with all Chile arrayed below him, "from the Pacific to the Andes," it really was worth it.

As we trotted down the mountain that evening into the gathering fog, Josh and I talked about reaching the peak versus turning back just short. This had been a real question for most of the day; Josh was still hiking his way into shape, and the day had been short and the climb longer than we had anticipated. Faced with a choice to turn back immediately or reach the summit but face a hike down in the darkness, we both chose to deal with the dark. We needed that triumph.

As we walked down, the darkness that had seemed like such a concern on the way up suddenly didn't bother me at all. And Josh appeared to be elated. He distilled it perfectly as we reached the bottom. "It's a lot better for jealousy with my former coworkers to say I made it," he said. "As opposed to saying, 'Hey guys, I *almost* made it to the top of the highest mountain in the coast range.'"

17: AMOLANAS HACIENDA

Darwin Slept Here

But I have too deeply enjoyed the voyage, not to recommend any naturalist, although he must not expect to be so fortunate in his companions as I have been, to take all chances, and to start, on travels by land if possible, if otherwise on a long voyage. He may feel assured, he will meet with no difficulties or dangers (excepting in rare cases) nearly so bad as he beforehand anticipated.

—CONCLUSION TO THE VOYAGE OF THE BEAGLE

"I AM TIRED OF REPEATING the epithets barren & sterile," Darwin wrote as he continued his trip north, riding his horse through the desert as the Beagle surveyed up toward Peru and the Pacific crossing. "These words, however, as commonly used, are comparative. I have always applied them to the plains of Patagonia, yet the vegetation there possesses spiny bushes & some dry prickly grasses, which is luxuriant to anything to be seen here."

I read aloud to Josh as we rode the bus toward the mining town of Copiapó. "Barren and sterile," I said, looking out the window at the baked earth, cracked waterless ravines, and sparse, stick-like bushes.

"We'll see if we can help him and think of some new epithets," Josh said. He deliberated for a moment.

"If a cow died out there," he said, "it would be *parrilla* by evening."

We passed a solitary farmhouse. Heat waves reflected off the metal roof, and a mule chewed on some yellow weeds at the edge of a fence. "It *is* getting pretty bleak and dead now," Josh said. "If this was the U.S., they'd have a giant Smokey the Bear sign. Fire danger: extremely fucking high."

"The whole journey is a source of anxiety to see how fast you can cross the desert," Darwin declared. And while Josh and I anxiously suffered and sweated in seats on the sunny side of the bus, Darwin dealt with a far more serious problem: His horses had no food. At his campsites he found sticks for firewood but not fresh greenery. And some animals, it appeared, lived on the sticks. Darwin recorded with amazement that the mules used for hauling loads to and from the mines subsisted "on the *stumps* of the dry twigs of the bushes." His own horses, pickier or more sensible, didn't eat for nearly three days, until he arrived at a small hacienda called Potrero Seco in the Copiapó valley. "To all *appearance* however the horses were quite fresh," he wrote apologetically. "No one could have told they had not eaten for the last 55 hours."

The next day, cruising out of the valley in a small rental car, Josh and I passed a bridge-crossing sign at Potrero Seco, in the middle of an expansive vineyard. We pulled off the road and asked a worker in the field if there was still a hacienda around the area, but he looked at us like we were unhinged and said no, there's just the winemaking house at Cerro Blanco, a few miles up the road. So we got back in the car and made for our original destination, Amolanas, a hacienda at the end of the Copiapó valley where Darwin met an elderly gentleman named Don Benito Cruz. "I staid there the ensuing day & found him most hospitable & kind," Darwin wrote. "Indeed, I defy a traveler to do justice to the good nature with which strangers are received in this country."

Darwin had visited plenty of haciendas in his time, but rarely were they labeled on maps. Amolanas was. It appeared to be located in the narrow southern part of the valley, at the end of a twenty-mile dirt-road connector off the main highway. After a familiar hassle with the rental-car company, which involved the saleswoman never having heard of Amolanas ("Where?" she asked), then calling three of her friends to find out if the road still existed (yes, they said, although they differed in their assessments of its condition), we sailed out through the valley.

The scenery looked similar to the Coquimbo valley, but even drier. The river didn't trickle, it oozed. The mountains didn't have cactuses, or scrub brushes, and instead started to look almost dune-like, leaving a cheerful blue sky and beautiful green cultivated valley separated by a white-hot ribbon of hostile, sandy wasteland.

At roughly the right spot near the end of the valley, a dirt road branched off to the farms and houses of the "Amolanas Packing Co." The road wound through a few vineyards and emerged at the spillway for a massive gravel dam on the river that had created a sterile-looking lake. Heavy sediment had turned the water a still, opaque green. We wound away from the lake and into the mountains. The road turned chalky, and without air conditioning we faced the choice of breathing heavy dust or roasting ourselves.

The road got narrower and kept climbing. Soon we were hugging a cliff edge with the tires struggling for traction in the loose dirt. Dust swirled inside and outside the car. We passed the skeletal remains of a dump truck that had jumped a switchback and crashed down into the next one. Its engine had popped through the grill. A few minutes later, as we inched along next to a fairly sheer precipice, I looked at the road ahead and saw a reincarnation of the dead mining truck bearing down the hill toward us. "Hmm," I said.

Josh, already nervous from the lack of traction and the steep drop-off, looked up. "What?" he asked.

I pointed out the red metallic glint of the truck as it wriggled along the switchbacks and blind curves above us.

Josh wiped his hands on his forehead and dropped our map. There was barely room for our small sedan to grip the road, squeezed as it was between the drop-off and the sheer bluff rising up a foot from my window. "Don't look down," I advised Josh.

"I've been looking down for some time now," Josh said. "Do you think they have the same rule in Chile about whoever's uphill being the one to back up?"

"Probably not when mining trucks are involved," I suggested.

We rounded a bend and saw the truck blocking the entire road one hundred meters in front of us. It would be a long, hazardous job of negotiating switchbacks in reverse to get out of the canyon. I spied a spot where the road got wider, halfway between car and truck, and I gunned the engine, allowing us to perch jauntily on the uphill side of the road while the mining truck, its oversized tires hanging off the cliff edge, cleared us with six inches to spare. The driver gave us a dirty look. I waved back.

"I don't think my heart could take any more of that," Josh said, trembling. He switched to his radio voice. "As I wiped my sweaty palms, I wondered whether Eric understood the full direness of the situation."

"I thought I saw you wiping your palms," I said.

"I wasn't sure at first if the map was making them wet or if I was making it wet," he said. "But it was me."

"I suppose the nice thing about being on the driver's side is that I can't really concentrate on the cliff edge."

"I concentrated enough for both of us," Josh said.

A nagging thought I'd been chewing on for the last few miles, à la Darwin's horse gnawing its post, suddenly revealed itself.

"I don't think Darwin would have come this way," I said. "He never mentioned leaving the Copiapó valley, and we've clearly done that."

We had gained maybe one thousand feet in elevation. Darwin's description of Amolanas didn't include anything about hiking in the scorching desert.

"I think," I said, "I have a pretty good idea what's actually at the end of this road."

A few minutes of driving later, the road flattened out at the mountaintop. Three more mining trucks idled there, waiting to pick up loads of rock. I swung our car around, hopped out, and ran up to one of the trucks as the driver leaned out the window.

"Hey," I said. "Amolanas is a mine, isn't it?"

"Yes," he said. He looked at me with a strange expression, as if to say, "Obviously."

I ran back to the car. "Sorry, Josh. We didn't need to come up here after all. And now we're going to race back, quickly, before that mining truck gets in front of us."

We scurried off like roaches fleeing the light and just beat the mining truck to the narrow road. Half an hour later, back on the main highway, we stopped at the lake's edge and let all the dust out of the car. Then, brushing myself off, I walked into a roadside convenience store. "Is there an old hacienda around here, named Amolanas?" I asked as I paid for a bottle of water.

"Well," the man said, "all of this is Amolanas. It's all part of the hacienda now. You could ask over by the vineyard houses there."

Back on the dirt road we found a middle-aged man standing, watching us. By this point we'd driven through the entire town three times. I figured that in exchange for getting to watch us making fools of ourselves, the townspeople owed us some answers. Besides, no one seemed to be doing anything. We had passed the same man three times already, and he hadn't moved from the post where he lounged, shirtless and in flip-flops.

"Hi," I said, rolling down the window. "We're looking for the oldest part of the Amolanas hacienda. Do you know where that is?"

He rubbed his chin and thought for a moment. "Hang on," he said. He disappeared inside a small shack for a few moments and brought out an older man and a young girl, presumably his father and daughter. Gringo-watching: Fun for the whole family.

I repeated the question. "Oh," he said, "Yes. You need to go back to the main highway, down about 500 meters and then take a left. Ask about Pesenti Oviedo."

Back on the highway we drove back and forth to scout out the area before deciding that the old man meant a small driveway that led off the road and into a fenced-off residential area. As we idled in front of the gate, a truck pulled up next to us.

"Can we follow you in through the gate?" I asked.

"Yes," the driver said.

"Is the oldest part of the Amolanas hacienda here?" I asked.

"Yes," he said. "Just go straight."

So we did. At the end of the driveway loomed a Spanish colonial-style ranch house. Three women sat on the porch, and as we approached, the eldest woman stood up, smoothed her tan skirt, and came smiling to greet us.

"Is this the old Amolanas hacienda?" I asked.

"Yes," she said. She said she was the caretaker for the mansion, which was once owned by an Italian rancher family, the Oviedos. Now she kept the building maintained for historical interest, as it was one of the oldest houses in the Copiapó valley.

"We're looking for the place where Charles Darwin, the English naturalist, spent the night. Is this it?"

She waved at a mud-brick structure behind the house. "I think so," she said. "It's supposed he would have stayed here." The mansion, she said, dated to the 1920s, but the low adobe structure was much older. "Go on," she said. "Take a look inside."

The walls of the adobe building were cracking, and some of the dried mud had chipped away to reveal straw and dried grass. Rickety wooden posts held up roof beams that looked ready to collapse. I pushed past an incongruous truck camper shell resting against the deteriorating wooden door and walked through into the dark, cool tile room. A couple of barrels rested in the corner. The adobe building certainly looked like it had been around for 170 years. "You know, with Chagas' disease the three things they say to avoid are old houses, adobe, and rural areas," Josh said, entering the room behind me. "Check, check, check. God should have hung a sign outside, 'Get Chagas' disease here.'"

Chagas disease, an illness spread by absolutely filthy South American insects called assassin bugs, had become one of Josh's major preoccupations in the desert. My own preoccupations, perhaps incautiously, lay elsewhere. In all my South American travels, I had never seen an actual building where Darwin had spent the night. Now, near the end of my time in Chile, I was suddenly confronted with this crumbling adobe surprise. My imagination scrambled to get its bearings. What would this room have been for? Who would have been here? What would it have sounded like? Smelled like? Would it have been as cool, or as dark? Did it have the same tile floor? Was it actually the same building? I didn't know of any books that mentioned this place. So many Darwin sites had been visited by other people. Could this be a place where Darwin spent the night, a place of immense historical interest, and yet unbeknownst to anyone except me, Josh, and the old caretaker?

Should I offer to buy the land before letting anyone else know?

Darwin slept here. The thought took a minute to register. The jungles of Brazil, the beaches of Uruguay, the plains of Argentina, the mountains of Tierra del Fuego, the forests of Chiloé all flashed through my mind, all as glorious steps on the way to this little adobe hacienda in the middle of the Chilean desert. It was a strange place to end up and yet a perfect one—a remote place, surrounded by mountains. Our day's journey had been an entire voyage in miniature: literal ups and downs, being chased by mining trucks, led on by random strangers, braving the elements, and huddling in the car until the elements had passed. Without the frustrations and false leads and hours spent coughing up dust and looking for the right path to the summit, there'd be no satisfaction in reaching the finish line.

In the last few lines of *The Voyage of the Beagle*, Darwin offers a summary of his journey. He wanted to encourage fledgling naturalists—a category he no longer belonged to—to strike out immediately for parts unknown. "In a moral point of view," the effect of traveling "ought to be, to teach him good-humoured patience, freedom from selfishness, the habit of acting for himself, and of making the best of every thing," Darwin wrote. "Or in other words contentment."

It's an exhortation just as true today as it was in 1836. A journey—of any kind, whether a daytrip to a local park or a year-long backpacking tour of the world—teaches contentment. It teaches the value of friendship, the kindness of strangers, the beauty of nature, and the exaggerated dangers of playing chicken with mining trucks on narrow roads. Traveling connects us to the world and renews our capacity to wonder.

"It's great to actually see something like that after looking for so long," Josh said, as we left.

"Absolutely," I agreed, slowly emerging from the misty mental images of 1835. "Still, they ought to have another sign out front."

"What's that?" Josh said.

"Darwin Slept Here."

ACKNOWLEDGMENTS

Darwin noted near the end of *The Voyage of the Beagle* that it's fairly well impossible to run off and travel without receiving generous and unconditional aid from a host of people. That's as true now as it was in 1835.

My own voyage started with encouragement from my parents, who displayed total enthusiasm for the project and also had the patience not to wonder (out loud) when I was going to get a real job.

In Rio de Janeiro, I was incredibly fortunate on my first day of traveling to run into a fabulous forest guide, Jean Marx Muñiz Belvedere, whose knowledge of nature and inspired hill-hopping set the tone for the rest of my travels. In Salvador da Bahia, Silas Giron was considerate to a fault and a perfect illustration of Darwin's maxim about depending on the kindness of strangers. Thanks also to Julia Hack-Davie for the introduction and the advice.

In Argentina, I owe particular thanks to Port San Julian historian Pablo Walker, for good conversation and a tour through San Julian's rocky history. Natalie Prosser Goodall and Thomas Goodall, the ranchers at the end of the world, were most helpful in answering my questions about life in Ushuaia. Monica Silva provided great directions to get me to the top of the Cerro Tres Picos.

In Chile, Roberto Pardo and Oscar Urbina Valdes helped us to a fascinating day in the mines, all without any advance notice. I'm still amazed that they let us through the security gate. Emer Mencilla Díaz jumped right into the spirit of things to lead us on two wonderful, dripping tours of the Senda Darwin. And Don Carlos Aguilar Cardenas and Antonio Gia patiently related the complicated mythology of Chiloe, and repeated it all about four times until I managed to understand.

In England, I was fortunate to meet with one of the world's great Darwin scholars, Charles' great-grandson Richard Darwin Keynes, whose contribution to Darwin research, including published transcriptions of Darwin's diary and zoology notes, made this book possible. In Shrewsbury, I was hospitably received by Rich Woolland and his Shrewsbury mates—thanks Miles, Mozza, and Becky.

I was fortunate to wander with two of the very best traveling companions. In Uruguay, Buenos Aires, and England, I followed in the footsteps of the incomparable, indefatigable, indomitable, and best of all, probably incurable, Nathan Cooper. Thanks for the sheep, the driving, and the tour of the Worcester china factory, and it's your turn to visit me, mate. In Chile, I chucked rocks and downed pisco with Josh Braun, a true Darwin-style scientific generalist, a great thinker, and still the kindest person I know, even after three weeks of being dragged to rain-soaked Darwin sites.

I wrote most of this book while I was a student at the University of California Berkeley Graduate School of Journalism, and I owe a great deal to three teachers there: Cynthia Gorney, Michael Pollan, and most especially Russ Rymer, a musician of English prose and possessor of cosmic perspective.

I also am indebted to my J-school colleagues, a group that is WAY better than the journalism overlords deserve. In particular, thanks to Larry Santana for reading an early draft, Tim Lesle for geological advice and carpooling, and inspiring comrades Cynthia Dizikes, David Gelles, Malia Wollan, and Charles Robert Foster.

Outside the journalism school, I received encouragement, advice, and editing suggestions from Hannah Naughton and Pamm Higgins. My agent, Diana Finch, managed to steer this book into being with enthusiasm and patience, and plus, she had the perfect name for the job. My editor at The Overlook Press, David Shoemaker, did a thorough, thoughtful edit for which I am deeply grateful.

Thanks most of all to my two longtime wing-people: Sarah Healy, for inspirational cheerfulness and a fun few days of touring Darwin's houses, and Brendan Buhler, who hates this sort of thing but really has earned it a million times over.

One last anecdote. Back in England and nearing thirty, Darwin drew up a pro-and-con sheet on the subject of marriage. "Constant companion (& friend in old age) who will be interested in one," swung the argument; at the bottom of the page he scrawled, "Marry—Marry—Marry. QED." The pros won handily, as they did for me. And so, finally, thanks to my wife Hari, who at this point could write the book *More Than You Ever Wanted to Know About Young Charles Darwin*, but hasn't.

INDEX